PRAISE FOR *DANCING WITH BEES*

"*Dancing with Bees* is one of the most important and accessible and entertaining books I've ever read. Brigit has poured meticulous detail and research into her book, which has left me with even more respect for our precious bees than I ever thought possible. What's more, it's a touching, sensitive account of what makes us human and how we connect to the natural world. Everyone should read it."

— KATE BRADBURY, author of *Wildlife Gardening* and
The Bumblebee Flies Anyway

"We are handed a lens — light, bright, beautiful things come into focus. Brigit's flare for observation and description, passion for knowledge, and ease with communication involve us in adventuring through the looking glass to explore with her the intimate life of wild bees. Gently, this timely book reminds us that nature is in trouble and that we must all join the dance."

— SUE CLIFFORD and ANGELA KING,
founding directors, Common Ground

"Brigit Strawbridge Howard is an excellent pollinator of information. *Dancing with Bees* is a book teeming with love: for bees but also for the natural world as a whole and, by extension, for life itself. Everyone who cares about the future of our planet should read it."

— TOM COX, author of *21st-Century Yokel*

"A beautiful book and one that hums with good life. Brigit Strawbridge Howard came late to bees but began noticing them at a time when their going was being widely announced. Her attention has been clear-sighted but also loving. By looking closely at the hummers and the buzzers, she has begun to take in the whole of what Charles Darwin called the 'tangled bank' of life, where there are bees (and Brigit's winning descriptions will help you know them) and there are plants, and there are other pollinators and nectar-seekers, including *Homo sapiens*. No other insect — surely no other animal — has had such a long and life-giving relationship with humans. Bees may well have shaped our evolution; our continued well-being is certainly dependent on them. Bees have long been part of our consciousness and art, buzzing in parables and fables and ancient and modern poems made out of their industry and their organisation and their marvellous sweet products. All that is in this book: It is ambrosia."

— TIM DEE, author of *Landfill*

"*Dancing with Bees* is a brilliantly described journey of discovery of bees, trees, people, and places, imbued with a childlike wonderment. Learn about cuckoo bees, carder bees, bees that are not bees, the commonplace and the rare. It is never too late to reconnect with nature and rewild oneself."

— STEVEN FALK, author of
Field Guide to the Bees of Great Britain and Ireland

"While the plight of our overworked honeybees elicits much hand-wringing, the rest of Earth's splendorous apian diversity has remained unjustly obscure. In this winning tribute to our black-and-yellow fellows, Brigit Strawbridge Howard celebrates the virtues of dozens of less heralded, but no less crucial, wild species — mining bees, leaf-cutting bees, mason bees, cuckoo bees. Like a bee herself, Strawbridge Howard is at once pragmatic and whimsical, flitting lightly between practical advice for crafting a bee-friendly garden and wise digressions about our manipulative relationship with nature. By the end of *Dancing with Bees*, you'll wholeheartedly agree that these indispensable creatures should be extolled as 'our equals, not our minions.'"

— BEN GOLDFARB, author of *Eager*

"A joy-filled voyage of discovery through the wonderful world of bees."

— DAVE GOULSON, author of *Bee Quest* and *A Sting in the Tale*

"In this delightful book, Brigit Strawbridge Howard brings us into the fascinating and often overlooked world of bees. She introduces us to solitary nesting bees that lay their eggs in empty snail shells, cuckoo bees that make other bees take care of their eggs, and the amazing social lives of bumblebees and honeybees. Her curiosity and wonder at these small creatures are infectious and will inspire a greater appreciation of our natural world."

— NANCY J. HAYDEN, coauthor of *Farming on the Wild Side*

"I devoured this book as I would a jar of exquisite honey. I was as fascinated by it as I would be watching a hive of bees at work. I may read another nature book this year, but not a better one. Or a more important one. As is made so manifestly clear in these pages, we need our bees. Thank God, then, for Brigit Strawbridge Howard, our queen bee-advocate."

— JOHN LEWIS-STEMPEL, author of *Still Water* and *Meadowland*

"*Dancing with Bees* is a passionate hymn to nature, a joyful celebration not just of bees, but of the power of paying attention. Strawbridge Howard's rediscovery of the natural world is infused with a sense of wonder both irresistible and infectious. And the promise of this beautiful book is that if we take the trouble to notice our natural surroundings, we too can find a way to reconnect not just to nature, but to a deeper sense of ourselves."

— CAROLINE LUCAS, MP, former Green Party Leader

"Well written and researched, beautifully illustrated, and packed with natural history detail, *Dancing with Bees* is a book to start you off on a journey that could well become an obsession. Even if you are well versed in the ways of bees, you will still want to wrap yourself in the warmth of this charming book. Brigit Strawbridge Howard gently shows you all the things you may have been missing; you are about to enter a macro-world of wonder and delight. I absolutely loved this book. If, due to infirmity perhaps, I am ever unable to walk in the countryside, I can now go dancing with bees whenever I choose."

— DR. GEORGE MCGAVIN, president, Dorset Wildlife Trust; honorary research associate, Oxford University Museum of Natural History

"Brigit Strawbridge Howard leads us on a wistful pilgrimage of awakening into the world of bees who are among the most fascinating, charismatic, and important of insects. Written in an easy, accessible style without shying away from solid facts and beguiling detail, and beautifully illustrated by renowned Devon naturalist John Walters, Strawbridge Howard's book is the result of hundreds of hours of watching, listening, and learning in her garden and the wider countryside, wondering what the future might bring and how human excesses may be curbed."

— STUART ROBERTS, entomologist

"Hovering through Brigit Strawbridge Howard's remarkable encounters with bees, alighting on beautiful and often unexpected descriptions of bumblebees, miner bees, and even parasitizing cuckoo bees, one dips into a world most of us have forgotten. By leading us gently and discretely into the minutiae of nature, Brigit shows how rewarding it is to reconnect — how the world's tiniest beings can not only lift our spirits, but signal the way to a richer, wilder future."

— ISABELLA TREE, author of *Wilding*

"Sprinkled with moments of pathos, this exquisite book is the perfect introduction to the often neglected world of wild bees — and the beautiful plants with which they dance an ecosystem into life."

— HUGH WARWICK, author of *Linescapes* and *Hedgehog*

"*Dancing with Bees* is an antidote to the reality of modern life that's spent nose down in our smartphones while the wondrous stuff — nature — goes on all around us. Brigit Strawbridge Howard chronicles her own journey of reconnecting with the natural world with heartfelt eloquence. Her descriptions of the creatures, plants, and landscapes that populate her journey are made with the unabashed joy of someone for whom a veil has been lifted, revealing a world to be cherished but also in great need of our protection."

— MATTHEW WILSON, garden designer; author; panelist, BBC Radio 4 Gardeners' Question Time

Dancing with Bees

Dancing with Bees

a journey back to nature

BRIGIT STRAWBRIDGE HOWARD

CHELSEA GREEN PUBLISHING
White River Junction, Vermont
London, UK

Copyright © 2019 by Brigit Strawbridge Howard.
All rights reserved.

Illustrations copyright © 2019 by John Walters. http://johnwalters.co.uk

No part of this book may be transmitted or reproduced in any form by any means without permission in writing from the publisher.

Project Manager: Patricia Stone
Commissioning Editor: Shaun Chamberlin
Editor: Robin Dennis
Copy Editor: Eliani Torres
Proofreader: Angela Boyle
Indexer: Shana Milkie
Designer: Melissa Jacobson

Printed in the United States of America.
First printing July 2019.
10 9 8 7 6 5 4 3 2   19 20 21 22 23

Library of Congress Cataloging-in-Publication Data
Names: Strawbridge Howard, Brigit, author.
Title: Dancing with bees : a journey back to nature / Brigit Strawbridge Howard.
Description: White River Junction, Vermont : Chelsea Green Publishing, [2019]
   | Includes bibliographical references and index.
Identifiers: LCCN 2019014244 | ISBN 9781603588485 (hardcover)
   | ISBN 9781603588492 (ebook)
Subjects: LCSH: Bees. | Bee culture. | Honeybee. | Natural history.
Classification: LCC SF523 .S778 2019 | DDC 595.79/9 – dc23
LC record available at https://lccn.loc.gov/2019014244

Chelsea Green Publishing
85 North Main Street, Suite 120
White River Junction, VT 05001
(802) 295-6300
www.chelseagreen.com

MIX
Paper from
responsible sources
FSC
www.fsc.org   FSC® C008955

For my mother, Isabel,
who would have been so very proud
to hold this book in her hands

CONTENTS

Realisations

I was quite shocked the day I realised I knew more about the French Revolution than I did about our native trees. The thought stopped me, quite literally, in my tracks.

I was in my early forties at the time, and remember thinking, in my state of shock, that I was lamentably no more aware of life outside the bubble that was my world than the inner-city children I'd read about who don't know that milk comes from a cow, or that an acorn grows into an oak tree. In fact, it wasn't *quite* that bad, but I was alarmed nevertheless that I could not confidently name more than half a dozen of the trees I had just walked past on my way to work. What about the rest? Which was which? I tried frantically to remember the names of all the trees I did know, working my way mentally through the alphabet from 'ash' to 'yew' and attempting to visualise the bark, twigs, and leaves of any of them. It was a sobering exercise.

The shock for me was not so much that I was unable to name the trees, as you don't need to know the given name of something to love and appreciate it. Rather, I was shaken by the fact that I had stopped *noticing* them. And it wasn't just the trees I'd stopped noticing. My three-times-weekly walk to work took me up and over the Malvern Hills from West Malvern to Great Malvern, along well-trodden paths edged with wild flowers; past large expanses of tussocky grass, bare ground, and low-growing shrubs; through areas of sparse vegetation amongst ancient granite stone; above the tree line, and there, beneath a ceiling of vast, ever-changing skies. But I was so preoccupied with the chattering in my own mind, and getting to work on time, that I was oblivious to the abundant and diverse wildlife afforded by this wonderful mosaic habitat that surrounded me.

How had I fallen so out of touch with the natural world that I now noticed the changing seasons more by how many layers of clothing I needed to wear to keep me warm (or cool) than by how many leaves the trees were wearing? When had I stopped seeing what colour they were, where in the sky the sun was setting, and which wild flowers were blooming in the hedgerows?

What had happened to the little girl who yearned, with every cell in her body, that she might close her eyes one night and wake up the next morning in Moominvalley, where she could and would sit on the edge of a bridge and dangle her feet in the river whilst Snufkin piped in the spring, then look at all the new and exciting wild flowers through a real, grown-up magnifying glass with the Hemulen? Where had gone the slightly older child who used to dream of living with Laura, Jack, and Black Susan in their 'Little House in the Big Woods' of Wisconsin, all safe and tucked up in their trundle beds whilst the wind and the wolves howled through the night outside their windows? And where was the curious ten-year-old who would have given her right arm, not to mention a year's supply of sherbet dips and Black Jacks, to spend *just one day* in the shoes of the young naturalist Gerald Durrell? Did that little girl still exist? If so, I needed to find her.

I scanned back through the years, searching for clues, wondering if there had been some particular event or moment when the child who used to be me had quietly drifted away. Having rediscovered a perspective on the world that had somehow been lost to me for the past three decades, I was determined not to let it slip away again. I vowed to nurture this fragile thing – this reawakening, this precious treasure – to help it grow and become fully conscious once again, and to protect it from whatever ill wind had caused it to bury itself under the blankets of my psyche, where it had hibernated and hidden for all these years.

I am by nature a thinker and a problem-solver, so to better equip myself on this journey back to nature, I thought I might start by exploring how connections and relationships in general have a natural tendency to shift, change, and peter out. The thing about disconnection is that you don't necessarily notice it happening. There are countless occasions over a lifetime when we are required to make conscious but insignificant decisions to disconnect from something or someone – for example, when we switch off the radio or bring a telephone conversation to its goodbyes. But these simple, everyday decisions are not followed by long-term, or life-changing, consequences. Other conscious decisions, such as handing in your notice at work or ending a long-term relationship, are likely to bring about life-changing disconnections, though because of their nature, decisions like these will invariably have involved considerable forethought as to how their consequences might pan out. Neither of these was like my disconnection from nature.

But there are also other ways in which our relationships change. There are times when existing connections shift or are weakened, but not entirely severed, by the conscious decisions we make – for example, when a young person chooses to move away from home and relocate to another area. The reasons for such a decision will doubtless have been considered with some care, but the knock-on effects might not have been anticipated. Parents and their grown-up children may remain in touch via the phone or email, or with occasional visits, but their knowledge and understanding of what is going on in each other's lives will inevitably be lessened. The love or empathy parents and children feel for each other does not diminish, but the children's lives, once they have flown the nest, tend to follow exciting, new paths of their own, and their parents' influence is either greatly reduced or ceases to have any bearing whatsoever. Consequences of this kind are rarely taken into consideration, because we accept them as a given. Children leave home; this is the way of the world today. Was my disconnection from nature the way of the world today, too?

I was sure that the disconnection from nature I recognised so abruptly and acutely in my early forties had not been caused by any kind of deliberate severance or conscious decision on my part, so it must have crept up on me over a number of years, decades perhaps, without my noticing. It was as though I woke up one morning and no longer felt I knew the person I shared my life with, that we had somehow drifted in separate directions, moved on to new things, and, ultimately, 'fallen out of love'. This analogy falls somewhat short, because it involves two conscious human beings rather than one human being and the whole of the natural world. But it's a good place to start when you are trying to make sense of something so huge and seemingly unfathomable. At least it was for me.

It just so happened that these realisations came at the very same time I had started to become aware of bees. I had no idea then how big a part they would play in my journey back to nature, how they would help and guide me, and how much I would learn from being around them. But I would soon find out. I was about to fall in love all over again.

Bees. Where to begin? Given the enormity of our reliance upon bees as pollinators of our crops, it beggars belief that most of us know so little about them. Mention the word 'bee' to most people, and images of hives, beekeepers, and honey are the most likely things that will spring to mind. However, if you were to give the same people a sheet of paper and a box of coloured pencils, and ask them to draw you a bee, most would draw something shaped a little like a rugby ball with striped yellow, white, and black bands, to which they might attach a head, six legs, two antennae, and a pair or two of wings – something that looks, essentially, like a bumblebee rather than a honeybee.

But in actual fact, Planet Earth is home to at least twenty thousand different species of bee. This is quite a staggering figure, one which surprises most people when they first hear it (it certainly surprised me), especially if they have previously only been aware of the existence of honeybees and bumblebees. Even more surprising is the fact that of all these different species, only 9 are honeybees and around 250 are bumblebees. There are also around 500 so-called stingless bees. The rest are 'solitary' bees, and it is amongst these species that I have found many new friends. (When I say 'the rest are', I am slightly oversimplifying. But more on this later!)

Most of us are aware that bees are important pollinators, but far from being in awe of the fact that something so tiny is capable of achieving something so extraordinary as pollination, we tend instead to take this gift – or 'service', as it is so sadly referred to these days by economists – very much for granted. I use the word 'gift' with consideration and awareness of the fact that a gift is usually something that has been given with intent to a recipient. As bees and other pollinators go about their daily business of foraging for pollen and nectar, their aim is of course to collect as much as possible to take back to their nest to feed, or provide for, the next generation of their species. Bees are no more setting out to 'gift' us than they are setting out to pollinate the plants they visit, but the result, in my eyes, is one of the most wonderful gifts that nature bestows upon the human race, and one without which we simply would not survive.

Bees pollinate flowering plants, this we know. But how exactly do they achieve this? How on Earth does a bee, newly emerged from its cocoon or brood cell, recognise which flowers contain the best sources of pollen and nectar for it? How can it tell which flowers have already been 'worked' and which still contain rewards? Can it easily find the same plant again? How does it access a complexly structured flower? Once it does, how does it extract the pollen and the nectar and carry them back to its nests? Does the plant have any strategies of its own for making sure pollination takes place? What happens if there are no pollinators around when the flowers come into bloom? And how do social bees communicate information to one another? So many 'hows' – and these are only those that relate to bees' relationships with flowering plants.

There are many more bee questions to which I have searched, and continue to search, for answers. I lap up everything I can glean from books and online resources, but there is only so far you can go without a little help.

Fortunately, there are experts out there who go out of their way to share the invaluable knowledge they have built up over years of studying and working in this field. Many of these scientists go further by demystifying the science to make it a little more palatable for laymen like myself. I am hugely grateful for what I have learned, first-hand, from these people, whom I thank in the acknowledgements. Without their help, I would undoubtedly have stumbled at the first block.

As well as finding answers to some – but by no means all – of my bee questions, I have discovered, along the way, a wealth of further information. This research has led me to explore new and exciting topics such as lepidoptery (the study of moths and butterflies), botany, and pollination ecology, to mention but a few. These enticements, though, mean I am constantly diverted and distracted from my bee quest. I could do with a few more lifetimes to digest and assimilate all this new information, but as

I have just turned sixty, and parallel lives don't exist, I have decided to focus most of my time on the bees.

I also spend hours on end watching bees and other insects in our garden, on our allotment, and wherever I am out and about, so that I might learn more about them and their behaviour. By observing them coming in and out of their nests – following them from flower to flower, listening to the sounds they make, and photographing them – I have come to admire, respect, and, quite often, be astounded by them. It is often by looking closely at the photographs I've taken, by cropping and enlarging them on my computer, that I find the most interesting clues about their physiology and behaviour, especially when it comes to working out how different species go about their vital business of transferring pollen. In fact, I always warn friends who suggest going for a walk that they might like to think twice about having me as their companion, for I find it quite impossible to walk past anything small that moves or catches my attention without stopping to examine and admire it.

Bees are incredible in so many ways, that the information I share about them in this book will barely scratch the surface of their existence. I am not, and do not pretend to be, an authority on bees, or on anything else for that matter. My aim is to introduce you to some of the native species I have come to know and love best, acquaint you with their life cycles (which, although similar in many ways, still differ from species to species), and talk about some of the basic differences between honeybees, bumblebees, and solitary bees. But in sharing the stories of these bees, I also introduce you to some of the species which live alongside them, such as hoverflies, hawk-moths, and potter wasps, and tell you what I have learned about the threats these creatures all face, including those posed by insecticides, habitat loss, and climate change.

I include as much of what I have learned as possible, but leave out anything I have not fully understood. There is nothing more confusing than finding yourself on the receiving end of information dished out by someone who has only half understood whatever it is they are talking about. There are already many wonderful books, papers, and websites that you can go to for further, or more in-depth, scientific information should you choose. I list those I have found most helpful in the bibliography.

More than anything, my wish is to share with you aspects of bees' existence that have, for the past decade or so, filled me with ever-increasing wonder and delight, as I have immersed myself in watching, listening, and tuning in to them, alongside all the other wild and wonderful creatures and plants I have come across in my travels. By sharing the knowledge I have gained through spending more time with nature – together with the observations, understandings, and realisations this has bestowed on me – I also hope to inspire you to see bees in a new light, to want to find out more about them, to cherish them, and to welcome them into your world, or at least into your garden.

Finally, because one of the simplest and easiest ways to help bees is to make our gardens bee-friendly, along the way I note the flowering plants my husband and I grow, and the habitats we have created, to attract bees – as well as all the other bugs and beasties that live in or visit our garden and allotment, since this book is not just about bees. It did indeed begin its life as a 'bee book', or more specifically as a book aiming to raise awareness about bee decline and its many contributing causes, but just as my own personal journey of reconnecting with nature has taken on a life of its own, so, too, has my book. It has been gate-crashed by all manner of other creatures that I have met and fallen in love with, since I first put pen to paper nearly ten years ago.

I hope you enjoy walking alongside me on my journey, and that as you meet them, you, too, might fall a little in love with these extraordinary, magical little beings.

The Honey Trap

I have not always loved bees as I do now. Even when I was a child, it was butterflies, baby birds, frog spawn, snails, and newts that most grabbed my attention, never bees – apart from bumblebees, that is, which I loved from an early age for their bumbliness and the humming, buzzing sounds they made. It wasn't till many decades later, whilst I was running an environmental charity in 2006, that I first became aware of the magnitude of bee diversity and the wonders of bees' relationships with flowering plants.

The media were at that time reporting mysterious and extremely worrying happenings on the other side of the Atlantic. Vast numbers of honeybees were 'disappearing' in the United States, and the phenomenon – which thankfully appears to have since abated – became known as colony collapse disorder, or CCD. One of its key features was the complete disappearance of worker bees. These bees were heading out on foraging trips, leaving hives full of eggs, grubs, and honey, but not returning. No trace could be found of them; and their queens, left untended, died. For several years in a row, around 30 percent of hives in the United States were reported to have failed, and at the peak of CCD, over the winter of 2006–07, an alarming

60 percent failed. People began to talk of a 'bee apocalypse', and extra funding was poured into research and outreach programmes to save the honeybees.

Like many others, I followed the story of honeybee losses with increasing concern. Initially, I was mostly worried about the implications to the human food chain if the losses being reported in the news were to continue. However, as I began to do a little research of my own into what was going on, my concerns and focus switched from how the losses might affect us to concern for the honeybees themselves, and then other insect pollinators.

Planet Earth is home to some 352,000 described species of flowering plants. These are, in turn, pollinated by at least 350,000 described, and many additional undescribed, species of pollinating animals. Plants and their animal pollinators have been evolving together for millions of years, and whilst some flowers have become specialists, and adapted to coexist with specific animals, most are generalists and are visited by many different species. Birds and bats pollinate flowers, as do rodents, marsupials, and lizards. But the majority of pollinating animals are insects: wasps, hoverflies and other flies, butterflies, moths, ants, flower beetles, and, of course, bees.

Bees are known for the important role they play in pollinating the planet's flowering plants. Through this role, they help sustain major ecological and agricultural systems. Bumblebees are vital for the cross-pollination of tomatoes. A group of native North American solitary bees, the so-called squash bees, can take the lion's share of the credit for the production of most commercially grown squashes and pumpkins. And without managed honeybees, the vast California almond crops would fail.

I was shocked when I realised the enormous scale of commercial beekeeping in North America, and horrified when I learned about the stresses these poor creatures are exposed to. The figures involved, and the distances the bees are transported, are mind-boggling. In 2017, migratory beekeepers shipped around 1.7 million honeybee colonies to and around California, where they pollinated 1.3 million acres of almond trees. These bees were in addition to the 500,000 colonies that were already resident in the almond valleys. The almond crop alone relies on trucking some 88 billion bees from their wintering homes, which, in some cases, are up to one thousand miles away. Back and forth go the hives on flatbed trucks, east and west, with stops to pollinate summer crops in the Midwest, before they finally get a rest over the winter. Early in the new year, they start the whole circuit all over again. Some of these hives will travel ten thousand miles of roads each year as the bees pollinate crops including apples, clover, canola, alfalfa, sunflowers, and blueberries.

Far from being surprised that honeybees were disappearing and dying from a mysterious 'disorder', I became increasingly surprised that any of them were surviving at all.

Our relationship with honeybees goes back at least nine thousand years, to the very dawn of agriculture. Archaeological evidence suggests that Neolithic farmers kept wild bees and gathered their honey and wax for medicines and food, but we have almost certainly been dancing with bees since long before then. A cave painting in Valencia, Spain, believed to be about fifteen thousand years old, very clearly depicts a human figure robbing honey from a hive high up on a cliff wall. Our human ancestors must have taken a liking to honey from bees before anyone ever attempted to manage them.

You may be under the impression that all the planet's honeybees live in hives, groups of which are known as *apiaries*. In fact, not all honeybees are managed by humans. Honeybees frequently set up home inside tree hollows, chimneys, wall cavities, roofs, and other spaces where they can and do survive, undisturbed, and unmanaged, for many years. Bees that swarm and establish colonies away from hives are often referred to as *feral*.

As it happens, I have a number of friends who happily share their homes with active honeybee colonies, more often than not, lodging in their roof spaces or chimneys. So long as the bees don't come inside the house, and whilst the weight of the honeycomb doesn't compromise the structure of the building in any way, such colonies are perfectly harmless. People often become quite attached to them and fondly refer to the colonies as 'our bees'.

But such colonies are not always welcome. I read a report recently about an NHS hospital in Cardiff that had become aware of a massive honeybee colony living in its roof space only after honey started seeping through the ceiling and down the walls into one of the wards. Needless to say, this caused quite a kerfuffle when it happened. Wherever they end up living, feral honeybees need a space large enough to accommodate a colony of perhaps sixty thousand bees or more, as well as a sufficient volume to build wax comb in which the queen can lay her eggs and the workers can store honey and pollen. Other than this, anything goes. Whether the cavity is tall and narrow, or long and horizontal, the colony will build their comb to fit the space available to them.

Honeybees that live in hives are often referred to as *domesticated*, though to my mind this term applies more to animals that have changed their wild behaviour to fit in with humans, than honeybees, which as far as we can see, have not changed their behaviour one bit to suit us. I believe the word 'managed' is far more appropriate – or, in the case of the large-scale beekeeping that occurs in the United States, perhaps it would be better to say 'farmed'. Either way, most of the bees kept in hives today are kept for the express purpose of pollinating human crops, or to harvest their wax and their honey, and often both.

Although some other bee species do make and store small amounts of honey, it is mainly *Apis* species that produce honey in sufficient quantities to make them attractive to beekeepers. It is honeybees' ability to store enough of this sugary food to see them through times of hardship, whether this be drought, heavy rain, or the cold of winter, that marks them out from most other species. This hoard, combined with their ability to adapt to use whatever space is available to them to build their homes, has been key to their success and survival not only in the wild but alongside humans as well.

But it is not just their honey, or their ability to pollinate crops, that has drawn generations of humans to keeping honeybees. There is something else, something resonating deep within the human psyche, that makes us want to care for them, to know more about them, and even to overcome our fear of them so that we might connect with them in some way. Perhaps it is their work ethic, or how everything they do is for the 'greater good' of the hive. Maybe it is because we aspire to be more like them.

I have been thinking about this a lot recently, for though it is the 'other bees' I have ended up championing, I also find myself wanting to know more about honeybees. From what little time I have spent with them, I completely understand why those who keep them grow to love them so much. I have a book, *The Life of the Bee*, written in 1901 by Maurice Maeterlinck, a Belgian poet, playwright, and beekeeper. It is without doubt one of the most beautiful and enchanting books I have read about honeybees. 'No living creature, not even man', writes Maeterlinck, 'has achieved in the centre of his sphere, what the bee has achieved in her own.' Maybe that's part of their attraction? That, if we get closer, we, too, might learn to achieve what the honeybee has achieved, that is, to be *truly* social, to cooperate and coexist as they do, not always putting ourselves first, but working together towards a common cause. Maeterlinck calls this 'The Spirit of the Beehive'.

Honeybees are highly developed social creatures, living together in colonies that comprise a queen, tens of thousands of female workers, and up to a few thousand male drones that are produced at specific times of the life cycle for the sole purpose of mating. In scientific terms, this level of social organisation, in which a single female produces the offspring, and non-reproductive members of the colony cooperate in caring for the young, is known as *eusociality*.

As well as living in communities with overlapping generations of adults and offspring, and cooperatively caring for the young, eusocial creatures are defined by the fact that they have a 'division of labour', with individuals assigned to specific castes, each caste having a clearly defined role that is not performed by others. Amongst honeybees, these are the queen, workers, and drones. Honeybee colonies and other eusocial units (including the five hundred or so species of 'stingless bees', called Meliponini, that are found in tropical or subtropical regions) are often described as *superorganisms*; each unit functioning as an organic whole and individuals within the unit being unable to survive by themselves for any length of time. A honeybee colony without its workers, or its queen, would be like a human body without its limbs, or its heart.

Eusocial species are capable of expressing extremely complex behaviours, including group decision-making. This complexity provides them with several advantages over their more solitary cousins. For starters, eusocial creatures are more efficient foragers, not only working together to find food and other resources but also communicating the whereabouts of their findings to other members of their colony. Because the queen does not need to care for her *brood* – that is, her eggs and grubs, or *larvae* – but can rely on a caste of thousands of workers to do so, honeybee colonies can grow very large. Their sheer numbers mean they are able to outcompete other insects for both territory and food. (This is why introducing vast numbers of honeybee hives to an area can have a negative effect on existing populations of native bees.) These enormous populations are also quickly able to build or repair their homes, or to mobilise to defend their hives and stores if they should come under attack. The advantages of sociality are balanced by disadvantages, such as the demand for large amounts of food to support the colonies.

There are other eusocial insects, including ants, termites, and certain wasp species, but amongst bees, true eusociality is the exception rather than the rule. Even bumblebees, though mostly social, fall just short of the more developed form of eusociality displayed by honeybees. Unlike honeybee colonies that can survive as a unit for many years, bumblebee colonies are *annual*, and until the first brood of worker bees hatch out, there is no worker caste in the nest to share the labour with the queen. Other bees display various degrees of sociality, including nest sharing and cooperative brood care, without being truly eusocial.

The majority of bees have no social traits whatsoever and are called *solitary*. Though solitary bees might live alongside each other, they usually have their own individual nests or nest entrances, and do not interact with others of their kind (unless, that is, they are mating). Whatever their social tendency, once you get to know a little about bees, you cannot help but see them in a different light.

Of the twenty thousand species of bees, it is the eusocial colonies of honeybees that are by far the easiest to observe, in large part because so many of them live in hives managed by beekeepers. I am not a beekeeper, but as it happens, my husband, Rob, *is* – and much of what I have learned about the ways of these bees has been through watching the comings and goings at the entrances to Rob's hives.

———

I met Rob in the summer of 2013, at a natural beekeeping convention, held at an outdoor conference centre in Worcestershire. It was organised by my friend, 'bee-centric' beekeeping advocate and teacher Phil Chandler. The convention had been arranged as an opportunity for beekeepers who prefer minimal intervention and a hands-off approach, to get together and swap ideas. Phil had invited me along to give a talk on 'other bees' – to fly the flag, so to speak, for our beautiful but less well known bumblebees and solitary bees. My talk was not scheduled till the Sunday afternoon, but Phil had invited me to join them for the whole weekend. I had for some time been curious to know more about natural beekeeping, so jumped at his offer. It was on the Friday evening, as people were arriving and setting up camp, and I was sitting with Phil's partner, Lesley, talking about how happy and content I was being single and living on my own, when I first noticed Rob. *He looks interesting*, I thought, as he walked past.

It's funny how much about a person you can read from the way they walk, how they carry themselves, and the clothes they are wearing, before even seeing their face or talking with them. My first impression of Rob, as he walked past, was that this guy was comfortable in his own skin, earthy (if there is such a thing), and in no hurry to get anywhere – all qualities I find attractive in a person. The other thing I noticed was that his hair was unkempt, a bit like mine. I guessed from his clothes and general demeanour that he was probably someone who worked outside. He certainly didn't have the look of someone who spent his life in an office. Even his tent was different. Instead of all the sensible, modern, lightweight tents everyone else had pitched, his was an ancient, French canvas tent, weighing at least five times as much as everyone else's and taking twice as long to put up. Maybe there was a bit of rebel in him, too? I hoped at some stage during the weekend we might get to meet.

And we did. Get to meet, that is, though not in a proper 'being introduced to each other' way, until the Sunday evening, when I was about to leave. But there were a number of occasions over the weekend where we 'connected'; I felt it, and he did, too. The first occasion was on the Friday evening, when Lesley and I were sitting by the campfire, eating our supper, and Rob just happened to be sitting on the bench next to us. He kind of joined in our conversation, but more by way of listening and agreeing with what we were saying, than contributing. His manner was confident, but very gentle and calming. I knew I liked him, and felt perhaps he liked me, too. We attended one or two of the same workshops over the weekend, and I noticed that whilst Rob listened intently to what was being said, he didn't put himself or his own views forward. He was there to learn, not to talk about himself, and he waited for others to ask questions before he asked any.

On the Saturday there was a dowsing workshop. We were all given divining rods made from old coat hangers and sent off to search for 'geopathic stress lines', above which bees are apparently more likely to build their colonies. I am ashamed to admit I found myself following Rob around the field. Dear me, I wince to think of it now. But Rob didn't seem to mind, and though we didn't find any bees, we did find an extremely healthy and active wasps' nest, right at the point where two geopathic stress lines met.

By the Sunday afternoon I knew for sure that I would like to see more of him, though I didn't yet even know his name. When I delivered my talk that day, I finished by describing a project I had set up in Cornwall to create habitat for wild bees. At the end I looked pointedly in Rob's direction and mentioned how desperately I needed volunteers. But it wasn't until I was saying my goodbyes at the end of the weekend that someone finally introduced us, and I discovered he was a gardener, that he lived near my parents in North Dorset, and, most important of all, that he was willing to come and do some gardening in my wild bee habitat in Cornwall. Three years later, in the same field where we met, we tied the knot.

Rob and I share many interests, most important being our love of nature. As I said, Rob is a gardener, but as we got to know each other, I learned he had been working full time in the same two-acre garden for more than a decade for Diana. Diana's garden is an absolute delight. A melding together of formal and wild, full of flowers, shrubs, trees, and wildlife. When he first showed me it, I understood straight away why he loved working there so much. I had never before seen so many different bees, or birds, visiting one garden as I had this one. As well as flower beds and lawns, there is an orchard, a vegetable garden, a small meadow, and two ponds, and all of it managed by Rob without the use of a single insecticide, herbicide, or fungicide. If ever you wanted to see proof that you don't need pesticides to produce riots of flowers and an abundance of fruits and vegetables, you would find that proof in this garden.

I feel blessed and fortunate that Rob's organic and nature-friendly ways mean that our own little patches of garden and our allotment in Shaftesbury, Dorset, are geared as much towards providing food for bees and other wildlife as they are to growing food for the table.

It is because of his love of nature that Rob's way of looking after his honeybees differs from that of many other beekeepers. He's been keeping bees for around ten years now, but very nearly changed his mind about being a beekeeper before he even started.

Rob's decision to keep bees came about when it dawned upon him how few honeybees he was seeing in Diana's garden. This suggested it was unlikely there were many honeybee colonies, managed or wild, already in the area. If there had been, they would certainly have found this garden. Honeybees can travel distances of more than three miles, though their natural foraging area tends to be within a couple miles of their hives. Had there already been large numbers of hives in the vicinity, Rob might have thought twice about introducing a hive to this garden, as he would not have wanted to knowingly add more competition to an area already saturated with honeybees.

Since the first reports of CCD and bee decline, beekeeping has increased enormously in popularity, especially in urban areas, where, as well as there being more individual beekeepers, there is a growing trend amongst well-meaning businesses to place hives on their roofs. But contrary to popular belief, honeybees do not need saving, and 'becoming a beekeeper' does not help 'save bees'. There are no fewer than twenty thousand different species of bee in the world, and the European honeybee is just one of them.

Whilst everyone has been focusing on honeybee decline, other bees have been quietly disappearing, or declining in range and numbers. There are currently nineteen native bee species on the United Kingdom Biodiversity Action Plan list, which is made up of priority species that have been identified as being the most threatened and therefore requiring conservation action. Of these nineteen species, six are bumblebees and the rest are solitary bees. European honeybees are not, and never have been, on the UK list. Nor are they considered to be an endangered species in North America.

In fact, honeybees are not even native to North America: They were introduced by European settlers in the early 1600s for honey production. Prior to this, all crops, including those that were brought in by the settlers, were pollinated entirely by native pollinators. So, yes, there are honeybee hive losses, and these losses are extremely worrying for beekeepers and consumers alike. Yet despite CCD and reports of high losses over winter, hive numbers globally have actually been increasing over the last fifty years.

Bringing hundreds of thousands more honeybees into an area where they might already be outcompeting native wild bees for foraging resources makes no sense whatsoever – unless, of course, you simultaneously plant fields full of flowering plants for the honeybees to forage upon, which, in most cases, is not what happens. Keeping bees might well help increase crop pollination, but the fact is that you are no more likely to save bees by becoming a beekeeper than you are going to save 'birds' by keeping chickens. Having said this, I have noticed that more and more beekeepers now take native bee populations into consideration before they expand their apiaries. The organisation Friends of the Bees specifically encourages this.

It is worth mentioning that not all the insects you see on flowers are actually pollinating those plants. I have discovered over the last few years that the terminology you use is extremely important when talking about bees and pollination. The term *pollinator* refers to creatures that actually pollinate the plants they visit, whilst all the others are called *visitors*. A bee or butterfly on a flower is not necessarily pollinating it. It might simply be supping nectar from the flower, without making contact with the plant's reproductive parts or getting pollen on itself to be transferred to another plant. This sounds quite obvious when you think about it, but it was not to me, until someone spelled it out.

Fortunately for Rob, Diana's garden was more than capable of sustaining a hive or two without it having any impact on the surrounding population of native bumblebees and solitary bees. In fact, he thought the garden would probably benefit from the introduction of some additional pollinators. And so it was that he duly signed up for a course with the local beekeeping association.

It is well known amongst beekeepers that 'if you ask ten beekeepers the same question, you will get eleven different answers.' This is no surprise. As no two people are alike, why would everyone follow the same ways when it comes to keeping bees? There is no doubt in my mind that pretty much every beekeeper, whether they keep bees as a hobby, a small business, or an industrial-scale agricultural concern, loves their bees. But when it comes to caring for, or 'managing', them, each beekeeper has a different view on how this should be done.

Rob tells me he learned a great deal from the course he attended, but being an organic gardener, he sought to model his beekeeping practices, and his hives, as much as possible on the way honeybee colonies live in the wild. Coincidentally, just before he completed the course, he received an email from his brother containing a link to an article by Phil Chandler about keeping bees in 'top bar hives' and asking him if he had come across such a thing. Rob hadn't, but as he had more or less decided he didn't want to keep his bees in a conventional hive, he looked them up.

Top bar hives are just one of many hive designs preferred by 'natural' bee-keepers. They allow the bees to build their own natural wax comb, involve less equipment than is needed to keep bees in more conventional hives, and can be managed with less intervention. As top bars are pretty basic in design and relatively easy to build, Rob set about making one himself, then collected his first swarm and began beekeeping. Rob currently has three top bars, each one containing a healthy, thriving colony. He has to date never lost a colony over winter.

As Rob's interest in the wild nature of honeybees has grown, he has begun to look at other hive designs, and in the summer of 2018 attended a course in Cornwall, where he made his first log hive. When wild honeybees set up home in tree cavities, they tend to site their nests four to six metres above the ground to give the colony protection from natural predators. Unfortunately, as we humans are so health- and safety-conscious these days, there are fewer hollow trees left standing than there used to be, which is where log hives come in. *Log hives* are basically hollowed out logs, fashioned from trees that have been blown down in storms, or felled when at risk of falling. They closely mimic the bees' natural nesting sites and can be strapped on a tree, at exactly the kind of height where honeybees would be likely to establish a new colony. Hives like these have been around for centuries, and are popular amongst beekeepers in some parts of the Continent. They have a very high success rate in attracting honeybee swarms.

Thanks to cabinetmaker and beekeeper Matt Somerville of Bee Kind Hives, who has pioneered the introduction of log hives to the United Kingdom, you can now easily acquire one – or learn how to make one yourself, as Rob did. Matt's log hives measure around fifty centimetres in diameter and eighty to ninety centimetres in length, with a hollowed-out cavity thirty centimetres in diameter. I ask Matt how important these dimensions are, and he tells me this is not an exact science, and he has had just as much success attracting bees with narrower cavities. Of course, this should have been obvious to me. After all, bees don't use tape measures. The thickness of the wood around the cavity helps to insulate the hive, reducing stress and energy consumption, maintaining warmth over winter whilst also helping to prevent overheating in hot weather. The hives also have a removable floor hatch for inspection, and a well-insulated roof.

We are planning to put our log hive up in a friend's tree this coming spring in hope that it might attract a wild swarm.

Until I met Rob, I had very little first-hand knowledge of any form of beekeeping, and I certainly knew nothing about the idea of keeping bees in the hollow of a log. However, I have come across a number of wild honeybee colonies in my travels, one of which I watched for seven consecutive years. I first came across this particular colony when I was out walking in the valley beneath West Malvern. I had stumbled my way down a very steep slope, and stopped at the bottom to take advantage of the shade offered by the largest of the valley's ancient oaks whilst my knees recovered. Here I became aware that a humming sound, which I had first noticed farther up the slope, was growing much louder, coming, I thought, from somewhere directly above me. I was blinded by the midday sun when I tried to look up, so moved round to the other side of the tree. The branches and foliage were more dense there, but once I had found a sight path up through the branches to where the buzzing was coming from, I saw them: hundreds and hundreds of honeybees, flying in and out of what must have been a hollow inside the tree. It was far too high up for me to see the entrance properly, but this was clearly a pretty huge and extremely active colony. I had seen bees coming and going from hives belonging to some friends, but nothing like this. This was like a dual carriageway of bees at rush hour.

People tell me that wild bee colonies are more susceptible to disease and parasites than managed colonies, but if this colony is anything to go on, that cannot always be true. I watched these bees throughout spring and summer over those seven years, right up until the year I left Malvern, and never once did I get a sense that they were unhealthy. I saw a swarm in a willow farther down the valley one year in May, and wondered at the time if it might have come from the old oak. I suspect it had. If only I had known then about log hives, I might have put one up somewhere in the area to see if one of the swarms from the oak tree might move in.

When a colony grows too large for the space it is inhabiting, it begins to raise some of the larvae as queens, and the old queen, together with up to two thirds of the adult bees in the colony, leave the hive en masse in a *swarm*. Honeybee swarms, if ever you are lucky enough to see them before they settle, are a wonder to behold. They remind me of starling murmurations, where hundreds, sometimes thousands, of individual birds move through the air as one, twisting and turning in unison till suddenly they drop down to roost. Honeybee swarms, too, drop down suddenly, often into a tree, though they may also settle temporarily on a hedge, around a lamppost, against the spokes of a bicycle, or in whatever random location they choose. There they form a giant mass of vibrating fuzziness, the thousands of worker bees clustered around their precious queen whilst scout bees go off to search for a new home. Once the swarm has decided on somewhere suitable to live, and moved in, they get straight to work building new wax comb, with hexagonal cells, where the workers can store pollen and honey, and the queen can start laying her eggs. And so, by the natural process of swarming, a single colony divides and becomes two.

Back in the old hive or tree hollow, a new virgin queen emerges and embarks on what is known as a *nuptial flight*. During this flight, she mates, mid-air, with as many as twenty males from other colonies. (It is believed she typically rejects the drones from her own colony.) She may make several such flights before settling down to her role in life, that is, laying eggs. Whilst there is a plentiful supply of pollen and nectar in the spring and summer, a queen may lay more than two thousand eggs per day. Her rate of egg-laying reduces towards autumn.

The population of any given colony, wild or managed, fluctuates greatly during the course of the year, according to weather, seasons, and available food resources. Whilst a colony can contain sixty thousand bees, more or less, at the height of the summer, there may be only a quarter of that number at the end of the winter.

Interestingly, though the normal life span for a worker bee is only around six weeks throughout the rest of the year, those that are produced from eggs laid in the autumn have more body fat and different metabolisms, which means they can live for up to five months, right through to March of the following year. When, at last, spring arrives, bringing with it warmer temperatures and longer daylight hours, those workers break away from their winter clusters and head out in search of food. Triggered by the workers' bringing the first of the new year's pollen back to the hive, the queen resumes laying once again.

The longevity of a honeybee queen varies greatly. If left to do her own thing and not replaced by beekeepers, or the colony, with a younger model, a queen can live for five to six years, during which time she might move home and set up a new colony twice or thrice. More often than not, however, she will survive for only a year or two, as some beekeepers routinely replace their queens. The male drones live for anywhere between a few weeks to a few months.

I had always assumed that the pollen brought back by worker honeybees was fed directly to the larvae, but I could not have been more wrong. Pollen, in its raw state, is indigestible. To make it digestible, the workers add nectar, together with saliva, gut enzymes, and wild yeasts. Over a few weeks, these cause the pollen to ferment. The resulting *bee bread* (also known as *ambrosia*) is eaten by nurse bees – worker bees whose specific job it is to care for the brood – to produce royal jelly, which, in turn, is fed to the queen and larvae, exclusively in the case of queen larvae and for about three days for others. So honeybees are true alchemists. Not only are they able to turn nectar into honey, but they have also perfected the art of breaking down the cell walls of pollen grains, by means of fermentation, to produce food for both themselves and their young.

The more I learn about bees, the more fascinated I become, but learning about bee bread has given me more reasons for concern. Now I have a grasp on how important the process of fermentation is in producing bee bread and larval food, and the role that wild yeasts play in that process, I can more easily understand how fungicides (which destroy wild yeasts), as well as insecticides, can cause the health of a colony to rapidly decline.

Understanding the threats to honeybees has helped me understand more about the threats to all other pollinators. Although scientists have still not uncovered the causes of CCD, or why it seems to have abated, lines of research during the crisis highlighted multiple driving factors, which have become known as the four Ps:

- parasites, with the main culprit in honeybee colonies being the *Varroa* mite
- pathogens, including diseases such as deformed wing virus, foul brood, and the fungi *Nosema apis* and *N. ceranae*
- poor nutrients – that is, a lack of plants and other matter for foraging
- pesticides, whether insecticides or herbicides and fungicides, which affect bees, their habitats, and their food

But despite the persistent media focus on honeybee losses, it is not only managed bees that face these and other threats, such as climate change. All pollinators do. The sooner we recognise this, the better.

Whatever its causes, CCD should have been a wake-up call. Bees are often referred to as 'canaries in the coal mine', and they were, and still are, telling us that something is wrong. We would do well to listen to them – starting with looking at how we have come to depend on managed honeybees for the pollination of so many of our crops.

Though I deplore the way in which commercially farmed honeybees are being exploited, I cannot think how the vast monoculture crops grown in North America would be pollinated if commercially managed honeybees were suddenly to be taken out of the equation. I wondered if perhaps more hives could be kept locally rather than relying on bees being trucked in from other states, but apparently this is not economically viable, as the commercial beekeepers would go out of business if their hives were called upon to pollinate only one or two crops each year. Also, it seems some of the regions where crops are pollinated by migratory hives would be too cold for honeybees to survive there over winter.

I wondered, too, whether the crops could not be pollinated by wild pollinating insects rather than by honeybees. And to a certain extent, this might be possible. However, of all pollinating insects, it is only honeybees that survive the winter en masse, ready to start pollinating very early in the year, when some big crops – almonds, for example – come into flower. So even if there were a way to revive native bee populations in the once biodiverse California valleys that are now dominated by almond orchards, these wild bees would not emerge early enough, or in sufficient numbers, to pollinate such vast acres of blossom.

There is, worryingly, no plan B. The world's growing reliance on industrial-scale monoculture crops, and in turn, upon honeybees to pollinate so many of these crops, has created a chicken-and-egg conundrum. It is only because honeybees *can* be managed and, equally important, because they are easily *transportable* that many crops grown on such a grand scale are able to exist in the first place. If we want to find ways to reduce the unnatural distances these bees and their hives are forced to travel and ways to improve their health by allowing them to live in habitats with more than a select handful of flowering monocrops, we need to rethink the whole system.

In an ideal world, we might return to a system where small-scale farming is the norm, but this cannot happen overnight, even if enough people demanded it. I hope and pray that the human race will somehow find a way to transition back towards a more ecologically sustainable, wildlife-friendly, and organic way of growing our food, but I fear this change might be forced upon us quite suddenly if we continue to remain blind to how broken and unsustainable our current system is.

Although we understand that bees pollinate flowering plants, we tend to take them for granted. Instead, we should be celebrating them, giving thanks every day of our lives, for pollination is nothing short of a miracle. It is well past time for us to recognise the gifts bees bestow upon us. And it is equally important that we recognise the contributions of non-bee pollinators, especially flies and hoverflies, without which global plant diversity would be dramatically reduced.

It took the apocalyptic headlines about colony collapse disorder to wake us up to the plight of the bees. Now that it seems to have abated, we must not be lulled into a false sense of security. Our pollinators need us. We need to step up to the mark and do whatever it takes – conserving what remains of their natural habitats, creating new habitats, and celebrating all the species of pollinators on our planet – so that they can increase once again in numbers and range.

CHAPTER 1

Spring on the Wing

Spring has sprung, and the world outside abounds with the sights, sounds, and happenings that herald its arrival. Bulbs are bursting into flower, hawthorn buds are tentatively coming into leaf, and tadpoles have already hatched out in the bucket that masquerades as a pond on our allotment. I haven't heard chiffchaffs yet, but our resident blackbird is singing his heart out from his favourite perch three-quarters of the way up the walnut tree.

The sun feels deliciously warm on my back as I potter around inside the greenhouse. I peel off some layers and start rummaging through the packets of bee-friendly seeds that came through the post earlier in the week. There are agastaches, salvias, and veronicas; also, mixed cosmos and wild mignonette. I think I might start by sowing the wild mignonette into plugs. There are older packets of cornflowers and sunflowers, too, that we didn't get around to sowing last autumn. They should still be viable, so I may as well sow them and see what happens.

Rob and I have a fairly good-sized allotment, but it is already bursting at the seams. Now is probably the right time to reclaim the area between the compost heap and the hedge, to make it ready for planting out whatever

I start off in the greenhouse today – unless, that is, Rob has it earmarked already for vegetables. I might as well get these packets of seeds sown, and if we run out of room, we can always sell the surplus on the plant stall we keep outside our house.

Before I go any further, however, I need to create some space under my feet. Rob is well ahead of me, so I am having to pick my way over his tubs and troughs full of spinach and winter salads. Our greenhouse is small and struggles to accommodate everything we ask of it.

Whilst I am shifting things around and mulling things over, Rob is outside planting our first and second 'earlies'. It is 17 March, St Patrick's Day; it is considered good luck to plant your potatoes today. We are trying some new varieties this year – Lady Christl and Apache – alongside old favourites such as Belle de Fontenay and Red Duke of York. We never bother with main crop potatoes, as we can buy these and most other roots from Liz at the Thursday market. But we love growing earlies and salad varieties, planting five of each in stacks of old tyres dotted around our plot on the allotment.

Ours is one of the plots at the very top of the allotment site, backed by a row of terraced cottages, most of them thatched, whose gardens run down to meet the allotment's boundary hedges. Each garden has a gate leading onto the allotment. Lucky them! We have come to know all the people who live in the cottages. There is a strong sense of the community in this part of our town.

Standing with the cottages behind us, our plot looks over the Blackmore Vale and down towards Melbury Beacon. There is a heavy mist hanging in the valley this morning, so I cannot see much farther than the farm below French Mill Lane, but the view is still spectacular. It is always so peaceful here; I especially love it at this time of year and feel blessed that we have been able to rent this plot from our local council. I wish everyone who wanted it, could have access to some land where they could grow and harvest food, or just sit and be. For some of our time in Shaftesbury, Rob and I lived in my mother's flat, which didn't have a garden. During those years, my mother's health began to deteriorate, and I struggled enormously to come to terms with the fact that she might soon no longer be with us. Being able to come and work, sit, or hide on our allotment was a godsend.

I am still deep in thought when I notice Rob waving his hands in the air to attract my attention. He is pointing at something on the ground. I poke my head out of the greenhouse. 'Bumblebee!' he calls. In my hurry to get there before she flies away, I trip over his spinach.

Nothing thrills me more than catching sight of my first spring bumblebee, recently emerged from her long winter sleep and preparing to establish a new colony of her own. Although one or two species have recently begun to raise broods during the winter months, most of the twenty-four bumblebee species found in Britain and Ireland have been hibernating deep beneath the soil since last autumn. I sorely miss their company whilst they sleep. It is such a treat to see and hear them again.

With bumblebees, it is only the fertilised queens that survive the winter. Apart from the odd colony active over the winter, all last year's males and female workers, together with the colonies' founding queens, will have died out long before the cold weather set in. So if you see an enormous bumblebee on the wing in early, mid, or late spring, she is highly likely to be a queen produced towards the end of last year's nesting cycle.

Having left her natal nest in autumn and mated with a male of the same species, a queen bumblebee spends the next few weeks of her life stocking up on nectar to help her build up her fat stores for hibernation. She then digs a tunnel into the soil, often beneath the roots of trees, or a north-facing bank, where she settles down for her long winter slumber. A hibernating queen can spend anywhere from six to nine months beneath the ground and can survive surprisingly low temperatures. Hard frosts and heavy snowfall pose no threat for her. If the temperature falls below a certain point, her body produces glycerol, a kind of antifreeze, which prevents her from freezing.

The first bumblebees to emerge from hibernation in Britain and Ireland are usually Buff-tailed bumblebees (*Bombus terrestris*), followed closely by the much smaller Early bumblebees (*B. pratorum*), then Tree bumblebees (*B. hypnorum*) and White-tailed bumblebees (*B. lucorum*). Others emerge, species by species, as their preferred food sources come into flower; some, like Great Yellow bumblebees (*B. distinguendus*) emerge as late as May or June, having spent the vast majority of their lives underground.

The bumblebee that Rob has spotted on our allotment is a beautiful Buff-tailed queen. She is sunning herself on the leaves of a dwarf comfrey that grows against the side of our compost bin, and she is *huge*. I always forget, until I see them again in spring, exactly how huge a queen bumblebee can be. Some species are naturally larger, but other factors come into play, too, for instance, how good a summer it was last year, or how well the queen was provided for by the colony's workers during her larval stage. I suspect this particular bee was extremely well provided for.

If she has only recently emerged from hibernation, our bumblebee will be peckish, to say the least. I look around to see what is in flower on our allotment. We have a large patch of lungwort, but unfortunately, it is of no use to our hungry queen. Buff-tailed bumblebee tongues are too short to reach the nectar hidden deep inside this particular flower. What else? The last of our hellebores are blooming, and there is a mahonia still flowering in one of the tubs. These are both good sources of pollen and nectar for Buff-tailed bumblebees, but I am acutely aware there is very little else in the way of early spring forage here, and there are certainly nowhere near enough plants in flower to provide fully for the needs of our queen. Rob and I really need to rectify this before next spring.

Looking up towards the cottage gardens, I can see hellebores growing in the garden nearest to us. There are also grape hyacinths and daffodils, but only in small patches. Our queen's best bet would be to fly up towards the town, where she would find the vast swathes of bright yellow celandine flowering on the slopes in St James's Park; better still, in the semi-wild area just beneath Park Walk, there are goat willows.

Willow catkins are magnets for bumblebees, honeybees, and other early pollinators, as they offer copious quantities of protein-rich pollen and carbohydrate-rich nectar at a time when very little else is in flower. Bees and other insects are essential to the pollination of willows. Only after the bees have pollinated the willow flowers will the seeds, like other catkin seeds, be caught by the wind and dispersed. (Interestingly, all the other catkin-producing trees in Britain and Ireland, including hazel, poplar, alder, and silver birch, are pollinated by the wind, rather than by insects.)

As if reading my mind, our beautiful bumblebee clumsily begins to take flight. After a bit of circling and lot of looping above our allotment, she heads off in the direction of St James's Park. I rather hope that her circling and looping were a sign that she was orienting herself, and she is planning to return here later. I would love for her to set up her home on our allotment, although, in truth, I feel she would be better off choosing a spot closer to the willows.

Ever hopeful that one of these house-hunting beauties might someday establish a nest in our garden, Rob and I have created a kingdom – albeit a Lilliputian kingdom – fit for a queen. We have been fairly limited in our efforts, because the property is rented and has but a tiny patio area, but our landlord has given us permission to pull up some of the flagstones and replace them with flowering plants, small shrubs, and a pond.

It is amazing what you can do with a small space, a few seeds and cuttings, and a little creative thinking. It doesn't take long to fashion a wildlife garden from scratch, and you certainly don't need to take out a mortgage to achieve it. If there are pre-existing wild or neglected areas on your plot, no matter how small or seemingly insignificant, all the better. Areas like these, even if they have not been long established, are already providing habitat capable of sustaining numerous invertebrates, garden birds, and other wildlife. All we need to do is nurture them.

We have such an area in the far left-hand corner of our patio garden, where an old Victorian privy has somehow survived. Its red-brick walls are covered in ivy that, when it flowers in the autumn, provides an abundance of nectar and pollen for numerous pollinating insects, as well as amazing nesting for garden birds such as wrens. The privy still has its original door, which doesn't quite shut any more, but it is dry enough inside, despite the tiled roof being in poor repair, for us to store the wood we burn in our stove through the winter. In addition to the inherited supply of well-seasoned wood left behind by the previous tenant, the privy is home to a variety of spiders and other creatures which greatly appreciate the shelter it affords. Robins nest there. Rats, too, I suspect.

Beside the privy grows a walnut tree. I call it the blackbird's tree. The idea of a walnut tree growing in a small patio garden must conjure up images of a grand old tree with a canopy so magnificent that, in the summer months, only dappled light might reach the patio beneath. Not so with the blackbird's tree. It is long and leggy and only just manages to hold its own between a twisted elder (pruned to within an inch of its life) and an extremely robust hawthorn. The three compete for what little light they can in an area dominated by our next-door neighbour's mature sycamores and, on the other side, by Sue's holly.

Sue is one of our neighbours, and we have to walk through her little plot to get to our patio. Very few of the gardens belonging to the terraced cottages we live in are attached to their respective cottages. Together with a row of Victorian privies, the gardens are all 'out the back', arranged higgledy-piggledy on each side of a narrow pathway, hidden from passers-by. These are secret gardens, where time stands still.

Sue's plot, my favourite, is home to an ancient, dilapidated tin shed that Rob covets above all other sheds. Its door, if ever it had one, is long gone, and it provides open and welcome refuge for all manner of creatures, great and small, putting the privies to shame. In high summer, the shed is almost completely obscured by honeysuckle, ivy, and wild clematis, which sprawl across the roof and climb through the holly tree that props the shed up. In fact, Sue's holly is probably the only thing that keeps the shed standing. The side of the shed still visible is flanked by foxgloves, evening primrose, and enchanter's nightshade. I have seen Silver Y moths nectaring on the evening primrose, and bumblebees find the foxgloves irresistible. Next to the doorway of the shed squat the remains of an old tree trunk, partly hollow and almost rotted down, a perfect microclimate for its jungle of mosses and liverworts. Ferns and fungi also grow here, in the narrow alleyway between the shed and the fence of our patio garden.

Sue uses the shed mostly to store logs, and she occasionally grows runner beans and courgettes against its south-facing side, but mostly she leaves it alone to do its own thing. This is the kind of wild shed that dreams and paintings are made on. And poems, too. Thank goodness she has no wish to tame it.

I wish more people were like Sue. I wish we weren't plagued by a seemingly insatiable and completely irrational desire to control, tidy, manage, and order everything around us. I know, at the end of the day, that Sue's shed is just a shed, a small tin structure sitting on a piece of slightly overgrown ground on the outskirts of a town in South West England. The fact that she leaves it 'be' is not going to save a species on the brink from extinction, nor will it stop the world from overheating.

But to me, Sue's shed and the little patch of ground it sits on are a celebration of humans living in harmony with nature, allowing our wild hearts to connect with wildlife, of our daring to leave something *alone*. 'Left-alone' places, whether they be old tin sheds, log piles, hedgerows, fields, or entire landscapes, are safe places for wildlife. Wherever there is minimal or zero intervention and management; where nature has the freedom to do what nature choses, rather than what we think nature should do; when we stop the clock, take a back seat, and become observers rather than masters; there, unexpected and magical things begin to happen.

My dearest wish is that Rob and I might one day be custodians of a small piece of land with a few trees, hedgerows, a meadow, and living water. It would be big enough for us to build a low-impact home so we could live there, and to grow food to eat and flowering plants to sell, but much of it we would just allow to go wild. This is our dream. In the meantime, I am just happy to have trees growing in, and adjacent to, our garden. Without them, there would be nowhere to hang our bird feeders and bee nesting boxes, and no cover for visiting or resident birds. I cannot imagine watching insects without there also being birdsong. The two go together, the birds and the bees.

When we moved into the cottage, it was immediately obvious that the spot in the corner of our patio, underneath the blackbird's walnut and the other trees, had been something of a dumping ground. It was piled high with broken terracotta tiles and pots and lengths of thorny bramble cut back by the previous tenant before he moved out. Black bags full of what I hoped might be rotted-down leaf mould turned out to be wet sand left over from when a new paving stone had recently been laid. An old tin bucket was filled to the brim with sludge. We left most of the broken terracotta and some of the brambles, as both were already providing good habitat for various creatures, but we cleared the rest.

This revealed the remains of a little wall, no more than forty centimetres or so high, made from Shaftesbury greenstone. There were more lumps of greenstone and plenty of broken bricks lying around, so we extended the little wall along the left-hand edge of the patio, towards the corner where we planned to dig our pond, gradually reducing the wall's height until it was only one brick high.

We ended up with just enough space between this edging and the hedge that marks the boundary between our garden and the one next door to put in some shade-loving plants for pollinators. We chose lungwort, geranium, dwarf comfrey, and hellebores, allowing each to spill out over the little wall onto the paving stones, then added foxgloves and columbine, for height, at the back. Rob dug out and lined a basin for the pond, and after it had filled with rainwater, we tackled the area behind it. This corner of the garden is slightly less shady, so we filled it with wild bergamot, yellow loosestrife, nepeta, and more foxgloves. We planted a pot of marsh marigold in the middle of the pond and, to soften the edge where it meets the patio, lots of bugle. If I were a bumblebee, I'd make my nest here.

After her long winter sleep, a newly emerged queen bumblebee, like the one Rob found on our allotment this morning, needs to forage for nectar to build up her strength and for pollen to develop her ovaries. Hopefully, if she is a species that comes out in the early spring, like our Buff-tailed bumblebee, she will have choosen a hibernation site close to an area with a plentiful supply of winter-flowering heathers, gorse, crocus, or pussy willow; or with other early spring favourites such as snowdrops, white dead-nettle, and green alkanet.

However, if the sun tricks her into emerging too early, and she can find nothing to feed upon, she will starve. It used to be that growing nectar- and pollen-rich plants that flowered from March through to October was sufficient to support our bees, but this is no longer the case. With changes in climate causing confusion about the right time to emerge for plants and insects alike, it is more important than ever to plant flowers, shrubs, and trees that will bloom in succession, all year round, including through the winter, in our gardens, parks, and other open spaces.

Once she has replenished herself with nectar and pollen, a queen bumblebee's behaviour changes. She begins to fly in a zigzag pattern, just above the ground, showing a particular interest where there are piles of dead leaves and rotten wood, as she prospects for a suitable site to build her nest. A bumblebee's preferred choice for a nest would be a vacated mouse or vole nest, but with the demise of hedgerows and woodland edges, these are becoming harder to find.

Other preferences (depending on the species) include tussocky grass, compost heaps, crevices beneath stone walls, bird boxes, or the eaves of houses. Those that are fortunate enough to find a suitable nest must be prepared to defend it from other bumblebees, as competition for nesting sites is great. I do not know for certain, but I imagine that one of the factors contributing to the success of Buff-tailed bumblebees and other early-emerging bees – whilst many other species are in decline – is that they steal a few weeks on other bees in the 'nest hunting' race, establishing their colonies well before later-emerging species can get a look in.

However, the main reason Buff-tailed bumblebees and other early-emerging species have been so successful is that they are not fussy when it comes to foraging and habitat preferences. Seven other species of bumble-bees are similarly flexible: the Early bumblebee, White-tailed bumblebee, Tree bumblebee, Red-tailed bumblebee (*B. lapidarius*), Common Carder bumblebee (*B. pascuorum*), Heath bumblebee (*B. jonellus*), and Garden bumblebee (*B. hortorum*). As these 'big eight' don't have to rely on only a few types of flowering plant or niche habitats for survival, they do very well in urban landscapes like parks and gardens.

Having found her supply of pollen- and nectar-rich flowers and stumbled upon a vacant rodent's nest or some other suitable abode, our new queen is now ready to establish her own colony. Depending on the species of bumble-bee, she might reach this important stage anytime between March and July, though an early spring can trick some bumblebees into starting earlier.

Before she begins to lay her eggs, she needs to remove any unwanted debris from inside her chosen home, and to waterproof her surroundings to the best of her ability. This task completed, her behaviour changes significantly again. Instead of zigzagging and meandering close to the ground, she begins to fly back and forth, to and from her nest, with great purpose, collecting pollen and nectar. On her return trips, the pollen baskets, or *corbiculae*, on her hind legs are absolutely laden with pollen, looking to all intents and purposes like saddlebags. On top of the nectar she carries in her honey crop, a bumblebee can carry up to 50 percent of her body weight in pollen.

Whenever you spy a queen bumblebee carrying large pollen loads, you can be pretty sure she has established, or is in the process of establishing, a nest. Inside the nest, she secretes slithers of grey-white wax from glands in her abdomen and uses these to fashion a little pot, about the size of a child's fingernail and shaped like Winnie-the-Pooh's honey jar. This she fills with her foraged nectar. Should she need to remain inside the nest for long periods of time due to inclement weather, the queen will use the nectar in this honeypot to feed herself and keep up her energy levels.

Once these housekeeping chores and preparation have been completed, the queen is ready to lay her first batch of eggs. She mixes together some of the pollen and nectar she has collected, with saliva, kneading the mixture into a little lump into which she lays between eight and sixteen eggs. From now until the day her first brood of workers emerge and are ready to fly, the queen bumblebee's time will be divided between 'brooding' and nipping out to forage.

Bumblebees 'brood' in much the same way as birds do. For the bumble-bee eggs to hatch successfully, the queen needs to sit on them and keep the temperature at around 30 degrees Celsius. She does this by disconnecting the flight muscles inside her thorax and shivering her muscles until her body reaches the required temperature. (This is also how bumblebees are able to heat themselves up in very cold weather, enabling them to fly when other bee species are unable, but they can do so only if they have consumed enough nectar to provide the energy necessary to vibrate their flight muscles.) Unlike most brooding birds, a queen bumblebee, when she first establishes her nest, is a single parent. She has no support, yet she still needs to make occasional trips to forage. Her foraging trips become short and sharp, to ensure the temperature of her eggs doesn't drop too low in her absence.

Once she has laid eggs, the queen faces the entrance of her nest, ready to ward off any unwelcome intruders. Her little nectar pot will be positioned close enough so that she can easily dip her tongue into it and sup nectar. This will keep up her energy whilst she is brooding.

Around four days after the queen lays her eggs, they hatch. For the next two weeks the developing larvae feed upon the pollen she has provided for them. As they feed, the larvae go through various growth stages, before spinning individual silken cocoons around themselves, inside which they will pupate. During pupation, some kind of cellular alchemy takes place – in much the same way as it does when caterpillars turn into butterflies – and two weeks later the bees emerge from their cocoons as beautiful, fully grown adults.

The first few broods in a bumblebee nest are always female worker bees. The workers are usually much smaller than their queen and will take on the roles of nursemaids, cleaners, guards, and foragers. After her colony is properly established, the queen rarely leaves the nest again. She now has workers to collect the pollen and nectar to feed subsequent broods, and her role turns exclusively to laying more batches of eggs.

Time ticks on, and as spring becomes summer the nest continues to expand and grow. The queen keeps on laying, whilst the workers, depend-ing on their size, adopt distinct roles within the colony. Larger workers, being capable of carrying more pollen and nectar back to the nest than their smaller sisters, usually take on the role of foragers, whilst smaller workers stay at home cleaning the nest, tending to the queen, and feeding the larvae. At the height of summer, if conditions are right and there has been a plentiful supply of pollen and nectar, the number of worker bees in a Buff-tailed bumblebee colony might exceed four hundred individuals. In other species, for instance, Early and Garden bumblebees, a nest will support significantly fewer workers, peaking at about a hundred.

If the nest is successful and the queen remains healthy, she will now be poised for the next, and most important, step in the nest's life cycle: the production of males and new queens. Up until this moment, the queen has laid only fertilised eggs, releasing a minute amount of sperm, stored inside her body since she mated the previous autumn, each time she lays an egg. These fertilised eggs contain her chromosomes and those of the male she mated with, and they all develop into female bees.

When the time is right, she switches from laying fertilised eggs to laying unfertilised eggs – that is, eggs that contain only her chromosomes – and these unfertilised eggs develop into male bumblebees. And as she starts to lay male eggs, the queen simultaneously switches off a pheromone she has been producing that has instructed her workers to raise eggs as worker bees. The switching-off of this pheromone, and the laying of unfertilised male eggs, is the beginning of the end of the colony's life cycle. Her final batch of fertilised eggs will develop into new queens.

After they have emerged from their cocoons and are ready to fly, the male bumblebees leave the nest, never to return. The new daughter queens follow soon afterwards. Interestingly, after she has switched off the pheromone that instructs workers to raise the eggs as worker bees, the queen loses her dominance over the workers, and some of them begin to lay unfertilised eggs of their own. The old queen, tired and weak, loses her grip on the colony, and chaos breaks out.

Whether or not the workers start laying eggs of their own, once the males and daughter queens have left, the colony is on borrowed time. Over the next few weeks, its inhabitants die off, and nature's detritivores – earthworms, dung beetles, millipedes, and the like – move in and clean up the abandoned nest. The new queens mate and go into hibernation; the males are dead before winter sets in. The entire future of the species now depends upon the newly mated queens surviving hibernation and successfully establishing their own nests the following spring.

The average life span of a nest is around eighteen weeks, but Early bumblebees complete theirs within twelve to fourteen weeks, meaning they are able to establish two, sometimes three, colonies in a year. Amongst this species, the new queens, if they emerge by early autumn, immediately start new nests, instead of going into hibernation. Common Carder bumblebees, in contrast, are very slow to establish their nests, and their colonies are still active well into the autumn, when most other bumblebees have completely finished their life cycles. If you see a bumblebee still collecting pollen in September, it is likely to be a Common Carder.

By chance, this spring, a Buff-tailed bumblebee chooses the newly created wildlife kingdom in our garden to establish her colony. I discover the nest on 20 April in the little Shaftesbury greenstone wall at the far end of our patio garden. How did I not notice them earlier? I must have walked past the nest's entrance at least a dozen times to fetch logs and hang out washing. No matter, I have found them now, and I am looking forward immensely to watching their comings and goings over the next few months. The fact that a queen bumblebee has chosen *our* greenstone wall over all the other greenstone walls in town (and Shaftesbury, let me tell you, is *full* of such walls) fills me with joy. It is the kind of joy you experience when a pair of blue tits nest in one of your bird boxes, or when you find tadpoles in a pond that you not long ago created.

I ring Rob, who has already this year found two nests in Diana's garden, to share the news, then celebrate with my first nettle tea of the year. Nothing tastes fresher or greener than tea made from young leaves gathered whilst the kettle is still boiling. Tea bags are fine, but they simply do not compare with tea made from leaves or flowers you have picked and infused yourself. I used to get stung when gathering the leaves, but that happens less frequently now that I have learned to be bold and 'grasp the nettle' firmly between my finger and thumb before plucking it away from the plant. I make the tea in my favourite mug – tall, thin, perfect for keeping my drink hot when I get distracted and forget to drink it – and take it back out to the garden with the last slice of the wondrously fragrant (and seriously delicious) lemon polenta cake Rob made for my birthday. I brush the dust and cobwebs off the garden chair that hangs from a nail in the old privy, and station myself right opposite the nest entrance, close enough that I can watch the bees come and go, but far enough away so as not to disrupt the flight path of the workers.

Almost as soon as I am seated, a bumblebee flies out of a gap between two of the stones. She doesn't pause to orient herself, so she must already have been back and forth a few times and already knows exactly where the nest is. Had she been new to the job of foraging, she would have gradually flown upwards, in ever-increasing circles and figures of eight, taking in the landmarks, near and far, that would help her find her way home. Bumblebees tend to forage within five kilometres of their nests, though distances as large as twenty kilometres have been recorded. It never ceases to amaze me that bumblebees are able to fly these distances yet still manage to find their way back home.

This particular bee is extremely small, even for a worker. Her diminutive size tells me she is probably from the colony's first brood of workers, and that the pollen the queen provided for this first brood might have been lacking in quality, quantity, or both. It is the pollen collected by bees that provides the protein needed for the larvae to develop and grow. Without a good source of protein, bees, just like any other animal, including humans, are likely to be stunted in size compared with those fed lots of good protein. Later broods in the nest's life cycle, which are provided for by multiple workers rather than the queen alone, tend to be larger – providing, of course, that good sources of pollen and nectar are available.

I monitor the activity at our bumblebee nest over the next month or so, but the first few weeks of June prove to be really busy, and I haven't caught up on it recently. The colony should be well established by mid June, but Rob tells me he thinks it is less busy than it was at the end of April; he hasn't noticed any bees going in or out for about a week. This strikes me as odd. The weather has been good, and there is certainly no shortage of suitable forage in the vicinity.

I am working to a deadline, but I cannot concentrate after hearing Rob's report on the nest, so I nip out to check on it. I can see from the steps into the garden that something is wrong, and I am horrified when I get closer and find the entrance has been completely blocked by some kind of rubbery web. I am further dismayed when I notice a number of large, maggoty-looking grubs moving around inside the rubbery stuff. I clear the grubs and the substance they are encased in away from the nest entrance with a twig and examine them. Although I have never seen any before, I am pretty sure these are wax moth larvae. I know they can pose a threat to honeybee hives, but I am not sure if they bother bumblebee nests, too. I wait for a while to see if there is any bee activity in the nest. Nothing. I put my ear to the ground to listen, in the hope that I might hear buzzing or humming, but the earth beneath the stone wall is silent. I fear the worst.

A few days later, my fears are borne out. No bees. The nest has failed. I am gutted – and absolutely furious with the wax moth, for laying her eggs inside my bumblebee's nest. But who am I to choose the success of one species over another? The wax moth was simply doing what wax moths do, providing the best start in life for her offspring.

I wonder, had her offspring looked less like giant maggots, whether I might have felt more kindly towards them. My curiosity piqued, I look to see what else I can find out about them and discover we have two species of wax moth, plus a Bee moth, in Britain and Ireland. Unfortunately for bumblebees, the Bee moth (*Aphomia sociella*), lays her eggs in bumblebee nests, usually sneaking in after dark to avoid detection by the queen. Once the eggs have hatched, the larvae feed on the wax secretions which the queen has used to construct her nest. Worse still, the moth larvae will also feed on the bee larvae.

I am not sure if the moth who laid her eggs in the entrance of our bumblebee's nest also laid eggs in the heart of the colony. It doesn't matter. By laying them just inside the entrance, and surrounding them in a web of protective, rubbery fibres, the moth blocked the resident bees from leaving, or gaining access to, their own nest. Within days, the nest would have been a tomb, the entire colony starved to death.

The Bee moth is just one of many predators, parasites, diseases, and other challenges that bumblebees have to contend with, which makes it all the more important that we do everything we possibly can to help them succeed. Fortunately for our fuzzy friends, they have their very own charity, a wonderful organisation called the Bumblebee Conservation Trust, or BBCT, set up in 2006 by Professor Dave Goulson and Dr Ben Darvill.

Back in 2006, Dave Goulson was based at the University of Stirling, where he had been studying bumblebees and the causes of their decline for over a decade. He and his research group had had numerous scientific papers published on bumblebees, but every one, says Dave, was 'read by a handful of other academics and then swiftly forgotten.' They had a pretty good idea why bumblebees were in trouble, but no one was doing much to help. It was extremely frustrating. Then Dave hit upon the idea of forming a membership-based charity 'devoted to providing sound advice by taking the best scientific research and turning it into real-world action.' And so the BBCT was born.

From its conception, the BBCT has aimed to inspire people to create flower-rich habitats for our beloved bumblebees. Crucially, as well as informing us gardeners, the trust provides a wealth of information and support for landowners who want to do their bit, and a dedicated army of members and volunteers raise funds and awareness in the effort to protect these iconic insects. Thanks to the BBCT and Buglife (another favourite charity), as well as campaigns by Friends of the Earth, the Royal Society for the Protection of Birds (RSPB), the Wildlife Trusts, and others, our wild bees and other insects are finally receiving the help and recognition they so desperately deserve and need.

I have, thankfully, had the chance to watch many bumblebee nests complete their life cycle, ending in the production of the new queens and males that ensure the survival of the species. And though the nest in our patio wall is sadly doomed, there are at least three thriving colonies, that I know of, on and around our allotment this year. Two of these – one an Early bumblebee colony, the other a Buff-tailed colony – are already well established and producing males. The third, a Common Carder colony, is a little further behind. The three colonies differ in size and character. Had I enough time to record all the comings and goings of each nest over a period of time, I might be able to hazard a guess as to their respective populations and where, exactly, they are in the life cycle. As it is, I am only able to label them as 'extremely busy and bursting at the seams', 'not so busy but still successful', and 'busy-ish, but quite laid back'.

The Buff-tailed nest under the compost heap is by far the busiest; indeed, at times, it seems the workers are queued up at the entrance, waiting to get in to deposit their enormous pollen loads. Some of the workers are so large that I could easily mistake them for queens, whilst others are no larger than my little fingernail. I would love to think that the foundress of this nest might be the queen Rob found sunning herself on the dwarf comfrey by our compost bin earlier this year, but this is wishful thinking. Either way, I am happy to see this colony is thriving on our plot.

The Early bumblebee nest, which I found underneath some leaf litter, is less busy. By that, I don't mean its workers are less busy, but rather there are fewer of them. Early bumblebees are the smallest bumblebees in Britain and Ireland, and this colony's workers are all quite diminutive. Yet, despite their size disadvantage, they were the first to produce males of all three colonies, and I am almost 100 percent sure I saw a new queen coming out of the nest last week. I would love for her to establish her own nest nearby.

The Common Carder nest is as busy and active as the others, but somehow seems more relaxed. I have spent more time watching them than the others. Their nest is located directly underneath the water trough where everyone on the allotment fills their watering cans. This communal trough is situated right on the corner of our plot, so I notice the happenings at this nest several times each day. The nest's entrance is at the front of the trough, beside the very patch of grass that most people stand on whilst filling their watering cans. I am worried the bees will be trampled on, and have told as many people as I can to watch out for them, but still I have to keep clearing their entrance where it has become clogged up with compacted soil and grass after heavy-booted feet have trodden on it. The bumblebees, bless them, seem completely unperturbed by this.

But there was one day, earlier this week, when I found three or four of them crawling around the grass, trying to find their way back into the blocked-up tunnel. I don't know how long they had been there, but as I cleared away the grass, they crawled right over my fingers and down into the nest to deposit their pollen. No fuss and no bother. Common Carders are, in my experience, the most gentle of all our bumblebees. I have a bit of a soft spot for this nest's residents.

I feel privileged to have witnessed the comings and goings of these bees and many more on our allotment and in our garden. More than that, I am glad to be *aware* of them. It amazes me that I spent so many years of my life not noticing the bumblebees right underneath my nose. I guess it must be down to some kind of 'selective seeing', a bit like selective hearing – where you hear only what you want to hear – only in this case, you see only what you want, or what you perceive you 'need' to see. I can't turn the clock back, but I am committed to doing more than my fair share of seeing and hearing now.

CHAPTER 2

A Nest of One's Own

You never know when you embark on a journey where it might lead you, and who, or what, you might meet on the way. Had I not taken an interest in the headlines about honeybees and colony collapse disorder back in 2006, I might never have discovered solitary bees. Considering that the vast majority of the world's twenty thousand bee species are solitary, it is incredible that so many of us are unaware of their existence. Yet the more I have learned about this diverse group of insects, the more fascinated I have become with their world. I am late to the party, but making up for lost time now I'm here.

The term 'solitary bee' is a bit of a catch-all. It is frequently used to refer to the thousands of bee species that are not honeybees, bumblebees, or stingless bees, though many of the bees that get lumped into this bracket are not solitary at all. In fact, some display varying degrees of sociality. However, understanding the technical aspects of how and why different species are grouped together, and where exactly they fall in the sliding scale of 'sociality', is a little like trying to understand English grammar; there are rules, but almost all the rules have exceptions.

True solitary bees are just that: *solitary*. They do not have a caste system, nor do they share responsibility for each other's young. Indeed, there is no overlap of generations; the parents die before their brood hatch.

Bees are believed to have evolved from early solitary hunting wasps, and so it is perhaps then not surprising that so many bee species are considered to be solitary. One of the fundamental differences between bees and wasps is that bees feed their larvae a vegetarian diet of pollen and nectar, whereas almost all wasps feed theirs a diet of prey (sometimes live, sometimes dead).

The wasp ancestors of bees were no exception to this rule. It is thought that at some stage, a wasp, probably from the Sphecidae family, inadvertently provisioned its nest with pollen, perhaps by bringing back prey caught inside flowering plants. As pollen contains protein, some larvae would have been able to develop solely on the pollen – and got a taste for it. Of course, wasps and bees don't 'think' about food as we do, but if they did, I can imagine some of those primitive solitary wasps thinking, *Wow, the grubs seem to be doing well on this yellow stuff, and it sure is easier than collecting food that fights back. Maybe it's time to switch to a vegetarian diet.* And so the first solitary bees evolved.

Whilst some solitary species make their individual nests alongside others of their kind and thus may appear to all intents and purposes to be living together in communities, they are not. These are not social or eusocial 'colonies' but nesting 'aggregations' of solitary bees, which can vary in size from just a few dozen up to hundreds of thousands of individual nests. Such aggregations can extend across areas of thousands of square metres, and if conditions are right, and food plentiful, may persist for decades. Aggregations like this are known as *bee cities*.

The first time I came across a nesting aggregation of solitary bees was when I lived in West Malvern, though I did not immediately recognise it for what it was. I had been for a long walk in the woods, and was resting on a bench beside the children's play park just above the valley where the wild honeybees live. The bench was cold. It was one of those old metal benches that wraps round a tree trunk, and as I sat there, I pondered what might happen when the trunk grew sufficiently in girth that it met the bench. Would the bench give way? I doubted it. I had seen many metal structures and barbed wires embedded in tree trunks before, and the trees seemed to survive, but I was not sure this tree would cope with being strangled by a metal bench. Hopefully, I thought, someone would notice the coming collision and remove the bench, so that it didn't ever cut into the tree.

I looked down to see if the bench had been fixed to the ground, and that's when I saw the bees. I noticed only one or two to start. Then I realised there were dozens. Some were very small, and buzzing around the ground too quickly for me to get a fix on them; others were slightly larger, crawling in and out of tiny individual holes, each about three millimetres in diameter, in the ground. Most of the holes were drilled into the bare soil around the bench, which had been compacted by years of people walking around the tree in order to take a rest on the bench. Some were circled by piles of freshly excavated soil, like miniature molehills.

I became aware that one of the larger bees seemed to be searching for something, its burrow perhaps? It kept flying up, circling a bit in the air, then back down again to the ground, where it pushed its way underneath dried leaves and little twigs, digging a bit here and a bit there, before starting the whole process over again. I was riveted by it. It took me a while to realise that my foot might be in the bee's way. As soon as this occurred to me, I stood up and picked my way (very carefully, so I didn't tread on any of the bees) back to the longer grass.

I was correct. My foot had been obscuring what was presumably the entrance to the bee's burrow. Worst still, I had squashed the earth with my feet so the poor creature was now having difficulty finding the entrance. Once it succeeded, it crawled quickly down into the earth and was gone. Most of the bees were flying far too quickly in and out of their burrows for me to get a proper look at them, but one of them stopped for a moment or two on a dry leaf before going into its burrow, so I managed to get a better look. 'It' was clearly a female: She was carrying bright yellow pollen on her back legs, and only female bees carry pollen. But the pollen wasn't at all like the smooth, tightly packed balls of pollen I had seen honeybees and bumblebees carrying on their back legs. This was more crumbly. And it was caked over a larger area of the bee's legs. Interesting.

At the time, I had only recently become aware of the existence of solitary bees. In the hope that I might attract some species of mason bee or leafcutter bee to nest in my garden, I had mounted some bee nesting boxes on the back wall of my patio just the previous week. I hadn't read that much about bees that lived in the ground (apart from bumblebees, that is), but as I watched these bees going in and out of their holes in the ground, it gradually dawned upon me that they might be a solitary bee species.

I continued to watch the mining bees for another half hour or so, before going home for lunch and scrolling through some of my bookmarked references online. Because of their appearance (densely covered in black hairs but with deep orange-red hairs on the thorax) and the fact that I had found them in February, they turned out to be relatively easy to identify. They were Clarke's Mining bees (*Andrena clarkella*), one of the earliest-emerging solitary bees in Britain and Ireland.

The pollen I had seen them carrying into their burrows was willow pollen, which these bees rely upon to provision their nests. The smaller bees, those that had been buzzing around above the ground but not entering the burrows, were males. I also learned that the stinging apparatus of *Andrena* bees is too feeble to penetrate human skin, so the mining bees in the park were of no danger to the children who play there.

When I returned to the same spot later that afternoon, armed with my newfound knowledge of the bees' behaviour, they were nowhere to be seen. I learned later that I would always be more likely to see these Clarke's Mining bees in the morning and early afternoon, because they usually stop flying once the sun has left their area in the shade. I continued to observe these bees, just as I watched the honeybees living in the old oak tree in the valley below, year after year, until I moved away from Malvern. They became one of my harbingers of spring, and I kept a diary of the date on which they first emerged each year. The earliest was 21 February 2011.

I have already mentioned the important roles that honeybees and bumblebees play as pollinators of crops and flowering plants; solitary bees are no less important. Like social bees, they collect pollen to feed their offspring, but the way they carry it back to their nests differs significantly.

Honeybees and bumblebees carry pollen in pollen baskets, and moisten it with saliva to ensure it is neatly and securely tamped down. As a result, they lose very little of this pollen en route. Once mixed with nectar, pollen becomes unviable and will not germinate, even if some of it does happen to brush off before the bees gets back home. The flowers visited by these bees must therefore rely on loose pollen clinging to their hairy bodies for pollination.

Solitary bees, in contrast, do not have pollen baskets and are far less fastidious. Apart from one or two species that carry pollen back in crops, the majority of solitary bees collect pollen in a *scopa*, comprising stiff, branched hairs located on their legs, sides, underneath their abdomen, and, in some species, on other parts of their bodies. Most female solitary bees pack the pollen in their scopae rather untidily, without tamping it down, or moistening it with nectar or saliva. This means pollen grains are far more likely to drop off when the bee visits the next flower, and, in turn, pollination is more likely to occur.

Added to this, most solitary bees are unable to carry as much pollen in each load as honeybees and bumblebees, so they need to make many more trips back and forth, to and from flowers and their nests. These extra foraging trips result in many more flowers getting pollinated. This messy method of collecting pollen is one of the reasons why a single Red Mason bee (*Osmia bicornis*), who collects pollen underneath her abdomen, can be around one hundred times more efficient at pollinating than a single honeybee. (Red Mason bees are especially important pollinators of apple crops.) So, you see, solitary bees are really the unsung heroes of pollination.

Solitary bees come in all shapes and sizes, from the minuscule *Perdita minima* of North America, which measures only two millimetres in length, to the world's largest bee, *Megachile pluto* of Indonesia, whose females can be up to thirty-eight millimetres in length and have a wingspan of nearly sixty-four millimetres.

They vary enormously in appearance, too. Some solitary bees are fluffy, some smooth, some rotund, some slim, some striped, some spotted. And they come in every colour of the rainbow, from the Box-headed Blood bee with its bright red, shiny abdomen to the enormous Violet Carpenter bee, so named because of its iridescent violet wings. There are dull black-grey bees that very few people ever notice, and bees, striped yellow and black, that look like wasps; and in the tropics, there are jewel-coloured orchid bees and shiny, metallic sweat bees, so bright and colourful that when you first see their images, you feel sure they must have been Photoshopped.

The diversity in their nest-building behaviour is equally astounding. Solitary bees can be broadly divided into two main groups: those that nest beneath the ground (often referred to as *mining* or *ground-nesting bees*) and those that nest in cavities above, and occasionally beneath, the ground (known as *aerial-* or *cavity-nesting bees*). The majority, like my Clarke's Mining bees, fall into the first group. The ways in which each species excavates, provisions, seals, and protects their nests can vary greatly.

Whilst ground-nesting species have evolved the physiological means and strength to excavate their own nests – hence the use of the name 'mining' bee – most of the cavity-nesting species tend to take advantage of whatever holes, tubes, and other cavities are already on offer. A few of the cavity-nesters, including carpenter bees such as *Xylocopa* and *Ceratina*, have become adept at burrowing into dead wood and hollow plant stems, and there are some, including the flower bees, that dig into soft mortar or clay.

However, most cavity-nesters care not a jot whether the main structure of their nest is made from natural or man-made materials, so will happily nest in a bee nesting box that has been manufactured specifically to provide them with accommodation. It is also not unusual to find them making nests in cavities such as garden hose pipes, wind chimes, keyholes, and watering can spouts. These bees are nature's ultimate opportunists.

Whatever a solitary bee's preferred nesting site, be it an aerial cavity or a burrow beneath the ground, their life cycles follow roughly the same pattern. With one or two exceptions (there are always exceptions!), male solitary bees emerge from their natal nests before the females, and spend a week or two either hanging around the nest area or patrolling flowering plants they know the females will forage on, whilst they wait for the females to emerge. Once mating has taken place, each female goes off on her own to construct and provision a new nest. The male plays no further part in the process. If you think of these bees as 'single mums', with each female taking sole responsibility for providing food and shelter for her own young, you'll begin to get the picture – though it's slightly more complex than that.

Unlike the aerial opportunists who take advantage of existing tunnels and cavities, ground-nesting solitary bees have to construct their own nests from scratch before they can begin to think of laying eggs. This is exactly what my first Clarke's Mining bee was doing all those years ago.

Different species nest in different locations. Some, like the Clarke's Mining bee, prefer to dig their burrows in level, compacted soil, such as that found in parks and on well-trodden pathways. Others excavate their nests in flower beds, lawns, cliffs, hillsides, sand dunes, even riverbanks – so long as the nest position is sunny enough, and the soil well drained, anywhere can be home. So if you want to create a space for ground-nesting bees in your garden, make sure you leave some areas of short grass for them. Best of all, if you have space, is to create a *bee bank* by piling up a mound of compacted soil in a sunny, south-facing position, not far from some wild flower lawn where they will find and enjoy the fabulous foraging you have provided.

Whatever location she chooses, a mining bee will begin the construction process by burrowing down into the earth using first her mouthparts, then her legs and body to create the main tunnel. After she has dug this tunnel, she will usually dig a number of lateral tunnels, each branching out from the main one. At the end of each of these lateral tunnels she fashions a small egg chamber large enough to accommodate a fully developed adult bee.

And this is where she gets clever. It is obvious that any nest constructed beneath the ground is going to be vulnerable to flooding. At the very least, it is likely to become damp. But ground-nesting bees solve any potential water problems by smearing the sides of their nest chambers, which are seldom made at the lowest point of the tunnel, with antifungal secretions. Depending on the species, these secretions are produced from either their salivary or abdominal glands.

No two species of mining bee are the same. Just as bees' body size, appearance, flight times, and food preferences differ, so, too, does their nest architecture. Some dig shallow burrows just one to five centimetres beneath the ground, but others burrow much deeper. The depth of the burrow depends in part on the soil type; bees that nest in very dry soil, for example, tend to dig deeper tunnels. In Britain and Ireland, solitary bees have been known to dig up to half a metre deep, but nests have been excavated in North America that extend to a depth of 2.5 metres. That's pretty impressive for a bee no bigger, in some cases, than a child's fingernail.

Structurally, mining bee nests can follow a fairly simple design, with a main tunnel running vertically down from the ground (or horizontally beneath the surface), then branching off in one or two directions, with a series of evenly spaced nest chambers along each branch. Others resemble labyrinths. I have become used to bees surprising me, but some of the diagrams of mining bee burrows in old encyclopedias have left me in awe that something so small is capable of constructing something so intricate.

Whatever the design, once she is satisfied that the nesting chamber is waterproofed, the female solitary bee sets about provisioning it with pollen, sometimes mixed with a little nectar, which she collects over many foraging trips back and forth to the nest. Most bees are not too fussy about which flowers they source their nectar from, but they can be slightly more fussy when it comes to choosing pollen. In fact, some, like Clarke's Mining bee, are such specialists that they visit only one plant for pollen – in their case, willow.

Once she has sufficient food for her offspring to feed upon, the female solitary bee lays a single egg inside each chamber, before sealing it with whatever building materials are to hand. More often than not this will be soil or sand. She repeats the process again and again, until she has laid anywhere up to twenty or so eggs. By the time she has finished laying all her eggs, she will have been on the wing for approximately four to eight weeks. Exhausted, and her job done, she dies.

Inside the nest, the eggs hatch into larvae, which feed on the provisions left by their mother, before their next stage of development, *pupation*. It is during pupation that larvae undergo a complete metamorphosis and emerge, at the end, as perfectly formed adult bees. After pupation most species of ground-nesting solitary bees in Britain and Ireland remain in their chambers, hibernating until the following spring or summer, when they finally emerge, mate, and start the cycle all over again. Later-flying species, however, tend to overwinter as full-grown larvae, and only pupate shortly before emerging. Some spend up to eleven months of their lives beneath the ground, but others, for example, the Yellow-legged Mining bee (*A. flavipes*), can produce two generations in the same year.

The life cycle of cavity-nesting bees is pretty much the same as that of mining bees. However, as these bees make their nests in cavities above ground, they do not have access inside their nests to excavated soil or sand with which to seal the individual egg cells or block the main entrance. This means they must forage for materials to protect their nest from both weather and raiders. Some species collect mud, sand, resin, tiny pebbles, or chewed-up leaf pulp from the surrounding area for this purpose, which gives them their name: 'mason' bees. Others, the 'leaf-cutter' bees, use cut leaves or flower petals. You can sometimes tell which species has made a nest simply by identifying the materials used.

It is easy to attract cavity-nesters to nest in your garden, and very much worth the effort. Not only do they provide endless hours of entertainment, but they also help to sustain local bee populations, in much the same way as planting bee-friendly pollen- and nectar-rich flowers does. Indeed, as more and more people have become interested in supporting native pollinators, garden centres and other outlets have responded by labelling certain seeds and plants as 'pollinator-friendly' and by selling bug 'hotels' and solitary bee nesting boxes. Whilst this is, on the one hand, wonderfully positive, it is important to note that many of these so-called pollinator-friendly plants have been grown in soil that has been treated with insecticides harmful to bees. So if you want to keep the bees that visit your garden out of harm's way, you might like to check the provenance of your plants before you buy them. Better still, grow your own.

Bee nesting boxes vary hugely in price and design. Some are fairly basic (holes drilled in blocks of wood), whilst others are extremely elaborate. Over the years, I have tried out loads of different commercially produced nesting boxes, as well as some that I have made myself. It has been interesting to note which have attracted bees and which have not. I had imagined that those made from natural materials might be the most popular, but this has not necessarily been the case. For instance, one of the bees that commonly nest in man-made bee hotels, the Red Mason bee, when offered a choice between a beautiful bundle of hollowed-out bramble, Himalayan balsam, and other plant stems, or a bunch of cardboard tubes lined with paper, seem to prefer the latter – at least, that is, in my garden.

Unfortunately, it has recently become evident that, far from helping bees, some of the nesting boxes people have been putting up – myself

included – might be causing more harm than good. If left unmanaged, many of these bee hotels will, in effect, become bee graveyards.

To find out more about this troubling phenomenon, I contact my friend Ron Rock, who has been experimenting with bee nesting boxes in his garden for a number of years. Ron is a volunteer for the Bumblebee Conservation Trust, but he has an equal passion for solitary bees, especially those in the Megachilidae family – that is, mason bees and leaf-cutter bees – which readily nest in man-made nesting boxes. Ron has been making his own nesting boxes for some time now, experimenting and adapting his designs year by year to provide optimum conditions for these bees. He has arrived at a design with a wooden frame, an overhang to protect the bees from inclement weather, and removable bamboo and cardboard tubes, about fifteen centimetres in length, lined with a roll of paper. So far this year, he tells me, around eight hundred tubes have been taken up, mostly by Red Mason bees. With each tube containing anything from one to thirteen cocoons, that's a lot of solitary bees.

His goal has been to make things as challenging as possible for parasites but as easy as possible for the bees, as well as the people who are managing the boxes. I use the word 'manage' here, but when it comes to caring for the bees themselves, Ron tells me he personally prefers the concept of *husbandry* – working *with* and caring *for* the bees as opposed to controlling their behaviour, which implies domination over nature. We chat a little about the semantics and agree that bees are our equals, not our minions.

I am interested to know more about how Ron defines bee husbandry. Is it enough to stick a bee hotel up on the wall and let nature do its own thing? 'In a word, no,' he says. 'If you make a conscious effort to attract bees to nest in your garden, I believe you have a duty of care to those bees. You certainly don't want your well-meaning efforts to inadvertently cause them more problems than they already have to contend with.'

This sounds very worrying. So I ask him how bee nesting boxes might make things *worse* for bees.

'Well, for starters, they are highly visible, and attract bees to nest in closer proximity and higher densities than they would in the wild, which means they are also likely to attract more parasites and predators,' Ron explains. Of particular concern are Houdini flies, Sharp-tailed bees, and woodpeckers. 'Houdini flies', Ron says, 'are actually *cleptoparasites*. They loiter around the nests of Red Mason bees and, whilst the unsuspecting females are out collecting pollen or mud, nip in and lay their eggs inside the cells alongside the bees' eggs, just before the females seal the cell. The Houdini fly grubs, when they hatch, eat the food left for the bee grubs, which means the bee grubs starve.'

It transpires that it is nigh on impossible whilst keeping a bee nesting box to avoid some of these parasites and predators, but it is at least possible, with good hygiene and husbandry, to keep them down to manageable numbers, and to minimise the worst of the problems. Ron does this by removing all the cocoons from his boxes at the end of each nesting season, and discarding any that have been parasitised. He also removes those affected by mould. By doing this, he gives the healthy cocoons a better chance of survival.

Putting solitary bee nesting boxes up in your garden *is* a good idea. They help increase local mason bee and leaf-cutter bee populations and provide hours of enjoyment, and because these bees are so docile, they are extremely safe for children. However, if you do decide to put up one or more, it is worth researching which are the best to buy or make, and which are the easiest to maintain. Also, don't forget to make sure you have enough flowering plants for the bees to forage on. If it is Red Mason bees you hope to attract, you will want to ensure there is a nearby source of damp mud. Should there be no rivers or streams in the vicinity, you can easily get past this obstacle by creating some nice muddy puddles in your garden.

I do so enjoy watching solitary bees going about their business, especially when I can see them building or provisioning their own nests. I have spent many a lazy afternoon in the garden watching female leaf-cutter bees snipping perfectly round or semicircular pieces of leaf from evening primrose, enchanter's nightshade, wisteria, and various roses before rolling them up, tucking up beneath them, and toting them back to their nests. There, they arrange them meticulously to line and seal the nests. The cutting process takes just seconds, but the lining and sealing take much longer. For a solitary bee lover, life doesn't get more exciting. Until, that is, you meet a snail-shell bee.

Just before I left Malvern for Dorset, I came across my first snail-shell bee, *Osmia spinulosa*. She was foraging on the viper's bugloss in my tiny patio garden on the side of the Malvern Hills, but so fleeting was her visit that if I hadn't had my camera with me, I might never have discovered what she was. By that time, I had begun to take photographs of pretty much anything and everything that visited my garden. I had already uploaded a few thousand images of two-, four-, six-, and eight-legged creatures to a folder on my laptop entitled 'Garden Visitors'.

Many of these visitors will remain forever unidentified, but when I enlarged the photograph I took of this particular bee, I thought straight away from her general appearance that she might be a member of the genus *Osmia* – a mason bee. She was not one of the mason bees I was familiar with, however, and she certainly wasn't one of my regular bee hotel boarders. Perhaps I'd get to add a new bee to the list of those I'd already recorded.

Suddenly, it dawned on me that I had been able to hazard a guess about this bee's genus *before* trawling for hours through the online image gallery kept by the Bees, Wasps and Ants Recording Society (BWARS). Too often my search would end without finding a visual match, and I would resort to posting my photograph on Twitter and asking if anyone knew what bee it was. I was tickled pink to realise that my bee-identifying skills had notched up another level. Given my abysmal track record in identifying solitary bees, the fact that I had managed to identify this one to genus level, entirely by myself, was worthy of celebration.

As it happens, I had spent a considerable amount of time that year observing several *Osmia* species that were using the bee boxes I had installed around my patio. I would watch, enthralled, as they carried little balls of mud or chewed-up leaf pulp into the various nesters. I couldn't see what they were doing inside the nesting tubes, but marvelled at the attention to detail they displayed whilst using mud, and leaf mastic, to plug up and seal the entrances after they'd completed their work inside.

Sometimes, after they had deposited a pollen load, they would crawl back out to the entrance, turn around, and reverse back into the nest. It took me a while to work out what they were doing, but once I did, it was obvious. They were laying their eggs. At other times, they arrived at the box with the undersides of their abdomens caked with so much bright yellow-orange pollen that I wondered how they had managed to take flight.

I glanced around and wondered where my new *Osmia* was making her nest. She wasn't using my nesting boxes, nor was she using the holes that had recently appeared between some of the bricks in the wall of my house. I needed some clues. Because each species has distinct nesting preferences, if I knew what sort of *Osmia* she was, I'd be able to narrow down my search.

It was relatively simple to pinpoint exactly which species she was, because of her distinctive blue-grey eyes. Many solitary bee species are impossible to tell apart without a microscope, but on this occasion I hit lucky. She was *O. spinulosa*, a Spined Mason bee. I scanned down to the description of this species' nesting behaviour and could barely contain my excitement. The Spined Mason bee, it seemed, makes her nest inside abandoned snail shells!

I was already amazed by the behaviour of these *Osmia* species, but to discover they have cousins that make their nests in snail shells was nothing short of a revelation.

The Spined Mason bee I spotted in my garden is one of just three snail-shell nesting bees in Britain and Ireland, the other two being the Gold-fringed Mason bee (*O. aurulenta*) and the Red-tailed Mason bee (*O. bicolor*). Each of these bees takes its own distinct tack when turning empty snail shells into nests.

The Spined Mason bee prefers small- to medium-sized snail shells, such as those belonging to the Brown-lipped, or -banded, snail. Having selected a suitable shell, the female deposits some pollen deep inside the shell and lays her first egg. She seals this cell with a partition wall using chewed-up leaves, then repeats the process with a second, and occasionally a third, egg, each following the other in a linear way, as though filling a curled-up tube. She is quite particular in her choice of leaves, favouring those from cinquefoil or burnet when preparing her nest.

Once she has finished laying her eggs, and partitioned off the uppermost cell, she seals the shell's entrance with a plug made from more leaf pulp. Finally, perhaps to give her offspring protection against predators and inclement weather, she crawls underneath the shell and cleverly manoeuvres it with her legs until the plugged-up entrance is facing firmly down towards the ground.

Depending on the availability of suitable shells, the Spined Mason bee might lay up to twenty eggs during her ten- to eleven-week flight period. Unusually for an *Osmia* species, for up to four weeks after she has completed each nest, the female regularly returns to her shells, making any repairs to the entrance plug that might be needed.

The Gold-fringed Mason bee – so called because of the beautiful, golden-coloured hairs on her thorax and abdominal segments – prefers coastal habitats. She lays her eggs in medium- to large-sized snail shells, seeking out those that have been bleached white by the sun. Their white surface will reflect the sun's heat and reduce the risk that the temperature inside the shell rises so high as to be lethal to her developing larvae. If she finds a suitably bleached shell abandoned by a Brown-lipped snail or White-lipped snail, she will lay three to five eggs in each. In larger shells, such as those of the Roman snail, she might lay up to ten eggs.

The Gold-fringed Mason bee builds her nest in a fashion similar to that of the Spined Mason bee, creating individual cells for each egg and sealing each with leaf pulp. However, where space affords, she often builds her brood cells side by side. Once she has filled the available space with brood cells, she turns her attention to the entrance plug, which she works at mostly from inside the shell, leaving a hole just big enough to climb out of before she applies the finishing touches to the outside of the plug.

Unlike the Spined Mason bee, the Gold-fringed Mason leaves her shell in situ, with the plugged-up entrance facing whichever way it was when she completed the plug, rather than turning it over, but this is not through lack of care or attention to detail. She has other tricks to protect her offspring. After the shell entrance has been plugged, she flies tirelessly back and forth, bringing more leaves to the shell. Then she uses the shell's surface as a sort of palette on which to prepare her pulp, leaving the impression, when she's finished, that the shell has been liberally splattered with dollops of bright green pesto. This also helps to camouflage the conspicuous white shell from potential predators, and may signal to other females that the shell is already occupied. I am filled with admiration for this little bee.

Yet, the nesting behaviour of the Spined Mason bee and the Gold-fringed Mason bee pale into insignificance when compared with that of the Red-tailed Mason bee, who is every bit as clever as she is beautiful. Of course, all bees are beautiful, but this species really stands out from the crowd. With her black eyes, head, and thorax, and the deep red-orange hairs on her abdomen, this aptly named bee looks as though she might once have been black all over, but has fallen into a pot of paint that has tinted the bottom parts of her legs and the whole of her abdomen. However, it is not this bee's beauty but rather her nesting behaviour that singles her out as remarkable.

The Red-tailed Mason bee chooses medium-sized snail shells for her nests, and to start with, things progress in much the same way as they do with her two snail-shell-nesting cousins. However, after laying four to five eggs deep inside her chosen shell and partitioning each egg off into a cell with some leaf pulp, the Red-tailed Mason bee goes to extraordinary lengths to protect her developing larvae.

First, she fortifies the remaining whorls inside the shell, filling them with a densely packed mixture of tiny pebbles, broken mollusc shells, soil, or chalk, which she has carried, or in some cases, dragged, back to the nest. When there is absolutely no space remaining inside the shell, she seals the entrance with a wall of leaf pulp. How on earth the next generation manages to dig their way through all this debris when they are ready to emerge as adults the following spring, I do not know. Still, there is now little way for a parasite or predator to reach her precious offspring.

Next she manoeuvres the shell so that its entrance faces the ground. This doesn't surprise me, because other snail-shell bees, including the Spined Mason bee, also do this. It's what comes afterwards that completely bowls me over.

Over the space of the next few hours, this bee diligently scours the surrounding landscape for pieces of dry grass and twigs – up to a hundred in total – which she grasps in her jaws and carries, one by one, back to her nest. The stems of grass and twigs vary in length, with some up to four times the length of her body. As she flies back to her nest, she looks for all the world as though she were a tiny witch riding a broomstick. I am enchanted. Nay, I am bewitched.

Back at the nest, she lays each piece of grass or twig very carefully up against the sides of the shell until she has created a kind of protective thatch or tent, which, when finished, completely obscures her shell. Unfortunately, her beautifully constructed thatch only stands until the first puff of wind arrives and blows it all away.

Red-tailed Mason bees live and nest in Shaftesbury, where I now live. I have watched them foraging on dandelions, willow, and ground ivy, traced their flights on their little broomsticks at the end of the nest-building process, and even discovered a completed nest. Frustratingly, I have yet to witness the nest-building itself. Maybe, if I collect lots of empty, medium-sized snail shells and hide them along the edges of our garden amongst a liberal planting of strawberry plants (whose leaves she likes to use to seal her cells); and if we grow all her favourite foraging flowers (which, according to Steven Falk's *Field Guide to the Bees of Great Britain and Ireland*, include sallows, blackthorn, and hawthorn, as well as low-growing flowers such as dandelions, violets, ground ivy, kidney vetch, and bird's-foot trefoil); then maybe, if I do all this, this beautiful bicoloured bee might grace us with her presence in our garden. I do hope so.

I adore the Red-tailed Mason bee. She fills me with awe and wonder, and embodies all that I love about the natural world. She is without doubt my favourite bee in the world.

CHAPTER 3

What's in a Name?

I was staggered when I first realised how many different species of bee we have living in Britain and Ireland, never mind around the globe. I wanted to know more about the beautiful and diverse world I had stumbled upon. Eager to get started, I borrowed a copy of Charles D. Michener's *The Bees of the World* from a friend, but as I scanned the list of names, I felt overwhelmed. Names such as *Lasioglossum calceatum* and *Anthophora quadrimaculata* seemed entirely appropriate for entomologists and other scientifically minded folk, but not, I thought at the time, for the likes of me.

My inability to pronounce the names of individual species, or to understand the meanings of the Latin and Greek words they stemmed from, left me feeling a bit intimidated.

For a while I managed to ignore the challenging names and bumbled along quite happily. I took photographs of the bees in my garden and saved them in folders on my computer using their common English names. My earliest folders, if I remember correctly, contained: Ashy Mining bees, male and female; a Buff-tailed bumblebee carrying an enormous pollen load; a Common Carder bumblebee, her head buried deep inside a yellow archangel flower; Hairy-footed Flower bees, numerous, of both sexes; a honeybee; leaf-cutter bees (at least two species, although I didn't know they were different species at the time) a solitary nomad bee, strikingly handsome; a Red-tailed bumblebee queen, quite enormous, foraging on a dandelion; Red Mason bees, a pair of them, mating; a Tawny Mining bee, female; and a Wool Carder bee (a lucky mid-air shot of a male approaching the lamb's ears on my patio).

I had painstakingly identified each of these bees using basic identification charts for the bumblebees and, for those solitary bees that were less common or more difficult to identify, the superb guide available on the BWARS website (which, I should point out, was never intended to be used for identification).

I say 'painstakingly' because my method of identifying each new bee involved quite a lot of pain. The BWARS gallery of photographs contains some 270 or so bee species, with multiple images of each species, including examples of males, females, and subspecies, as well as examples of queens, males, and worker bumblebees, so my searches required a considerable amount of both time and patience. It occasionally took me up to an entire day to identify one single bee. With some species I was able to find a visual match, but no common name, so I was stuck with scientific names I didn't understand and often couldn't pronounce.

Frustratingly, though I longed to discover the identity of these insects, the time I had available was extremely limited. I was then running a grassroots environmental charity, The Big Green Idea, which friends and I had set up to demonstrate how easy it is, if you have the inclination, to live a 'greener' lifestyle, regardless of your income. We'd established the charity shortly after I had appeared, together with my now ex-husband, Richard, and our children, James and Charlotte, in the 2006 BBC Two series *It's Not Easy Being Green*. Following the airing of the first series, we had received thousands of emails from people who thought it was easy enough for middle class or wealthy families, who could afford to buy wind turbines and organic food, to go green, but not for others.

Since many of the 'greenest' people I knew happened to live on very modest incomes, I was keen to find a way to dispel this misconception. With a team of volunteers, I visited schools and town centres around the country, travelling in a double-decker bus, which we planned, ultimately, to run on recycled vegetable oil. It was fitted out with a solar-powered cinema, educational displays, and a workshop area where we gave talks on everyday actions from recycling and composting through to making natural skincare and household cleaning products. We also hosted talks on 'big' environmental issues such as waste management, peak oil, deforestation, and climate change.

It was during this time that I first began to give short talks about bees and bee decline. Initially, I focused on the causes of bee decline, including colony collapse disorder, rather than on the bees themselves, but it soon became obvious that people were far more likely to do something to help if they knew a little about this charming group of insects. Whatever facts and figures somebody might dish out, whilst food is still available on supermarket shelves at a relatively cheap price, it is unlikely that talk of pesticides, parasites, pathogens, or poor foraging will lead to a mass shift in consciousness or behaviour.

So I stopped lecturing audiences about how important bees are as pollinators of human food crops and their monetary value to the economy, and began instead to share some of the photos from my garden, plus some of what I had picked up about the diversity, behaviour, and life cycles of various species of bees. My hope was to endear bees to my audience and inspire or motivate them to get out there and fill their gardens to the brim with pollen- and nectar-rich plants, to leave the dandelions in their lawns to flower, and to allow the bumblebee colony nesting under their compost heap to remain in place until their life cycles had come to completion for the year. Better still, someone might go home and appeal to their local council to adopt more bee-friendly practices in open spaces, or ask their MP to support more pollinator-friendly agricultural practices.

I looked through my photographs, choosing those I thought most likely to amaze and impress: the leaf-cutter bee carrying a deftly cut leaf back to her nest, the Buff-tailed bumblebee so loaded with pollen that it was a wonder she could even take off. I married these visuals to equally amazing and impressive stories of bee behaviour. Talking about the colourful character of Hairy-footed Flower bees would raise a few smiles. And I knew when I described how a queen bumblebee disconnects her wings from her flight muscles, then shivers those muscles to keep her eggs warm in her nest, my audience would see bees in a different light.

Above all, I wanted to change the overall emphasis of my talks from *head* to *heart*, by appealing to people's innate desire to protect something they know to be vulnerable or helpless, and to fight for something they care for. I wanted the people who learned about bees from me to leave the room *loving* bees, or at the very least wanting to know more about them.

As I was trying to bring more heart into my talks, I noticed how much easier it was to captivate an audience when I referred to bees using their vernacular names. But though I was able to wax lyrical about the velvety Red-tailed bumblebees foraging on my dandelions, the fox-coloured Tawny Mining bees nesting in my lawn, and the feisty Wool Carder bees patrolling my lamb's ears, what about the tiny metallic green bees I had found in an old wall near St Ann's Well? Or the large aggregation of ground-nesting bees with striped abdomens that had made a home in the municipal flower bed between Waitrose and the car park? Neither of these species had vernacular names at that time, so if I wanted to show off their photographs or talk about them, I had to master their scientific names.

I was also becoming increasingly aware that whenever I looked up a bee on the internet using the vernacular, my searches were not always productive. It was easy enough to find information about species that had been well researched, like the Buff-tailed bumblebee or solitary Red Mason bee, but quite a different matter when I attempted to search for more in-depth information on these bees without using their scientific names, or *nomenclature*. It was long past time for me to meet bees by their 'proper' names, including learning the meanings underlying the Greek and Latin. I tried one on for size. 'Hello, *Bombus terrestris*,' I said out loud, enjoying the sound of '*Bombus*' and intuiting that '*terrestris*' probably had something to do with the earth. So far so good. But how could I find out more?

The obvious answer was to consult my mother, who, at the grand old age of eighty-four, was then still teaching Latin. She would have been delighted to tutor me. Indeed, she would have relished having me acknowledge, finally, that I should have persevered beyond O level with this subject. However, knowing my mum, I could see exactly how my education would proceed: She would insist on first guiding me through the world of declensions and conjugations that I had so gladly left behind at the age of fifteen. She might even have tried to persuade me to join the U3A Latin classes she held around her dining room table on Wednesday and Friday afternoons.

No, this would not do. I wasn't interested in learning the language. I simply wanted to learn the names of the bees.

What I desperately needed was an illustrated guide to bees, one suitable for a beginner, but such a thing did not (at that time) exist. I was also woefully short of friends who had studied biology, entomology, or any other kind of 'ology', for that matter. So no help there, either. I would have to do this the hard way, on my own and from scratch.

I started by entering the word 'bees' in my search engine. Wikipedia defined bees as 'members of the vast insect order of Hymenoptera, which also include ants, sawflies and wasps.' If you, like me, do not have much in the way of a scientific background, you may not have stumbled across the word 'Hymenoptera' in your usual reading. It comes from the Ancient Greek words for 'membrane' (*hymen*) and 'wing' (*pteron*) and refers to the *taxonomic order* to which all bee species belong. So those Latin lessons with my mum would have left some gaps anyway. Not merely because she couldn't help me with the Greek, but because I also needed a better understanding of taxonomy.

Biological taxonomy is the system by which scientists name, describe, and classify all naturally occurring organisms. The term 'taxonomy' itself comes from two Ancient Greek stems – *taxis*, meaning 'arrangement', and *nomos*, meaning 'method', 'custom', or 'law'. So, taxonomy is a method of arrangement. Humans do so love to arrange things, sort them out, group them together, and organise them – sometimes literally, as in boxes, or virtually, as with computer folders. A good filing system creates order out of chaos, making it a great deal easier to find objects or information when we know where to look for them. It also makes it easier for us to share information with others. We do not have to be standing there, guiding someone to the correct drawer and folder in a filing cabinet.

Unlike some other forms of classification, biological taxonomy is designed to group together organisms that are biologically 'related' rather than just similar in appearance. Once you get the hang of it, it is no more intimidating than organising the cupboards in your kitchen. There are no rules dictating how you should organise your cupboards. You can arrange your food however you choose – according to sell-by date, frequency of use, or type of food – grouping together the tins and jam jars, the baking ingredients, the breakfast cereals with the teas and coffee, and so on, based on the fact that they tend to be used together. Typically, people organise their cupboards intuitively, deciding which things belong together before putting them away, sometimes revising the organisation as they discover that they rarely use that chamomile tea in the pretty box at the front of the cupboard.

Biological taxonomy differs from the kitchen cupboard example in that its groupings are mandatory, not flexible – though some organisms may get moved around as scientists discover new aspects of how they are related to other organisms. Every form of life must be placed in a series of ranks, starting at the highest level, domain, which is split into three categories: bacteria, archaea (single-celled organisms whose cell has no nucleus), and eukarya (everything else, including us). The ranks then descend kingdom, phylum, class, order, family, genus, and species.

Whilst the ranks used to classify organisms are relatively recent, the concept of biological taxonomy is not new. It dates back to the days of Aristotle, whose system of classification shaped the way naturalists recorded and described organisms for two thousand years. But the foundations of our current system of classification were laid down by the seventeenth-century English naturalist John Ray, who, in his efforts to catalogue and organise his findings, was the first person to define biologically what a 'species' means – that no matter what differences might appear in an individual organism compared with others of its species, it developed from the 'seed' of that species and is closer to it than to any other.

A century later, Ray's systems of classification were further refined by the Swedish botanist Carl Linnaeus, who collected and preserved model specimens of plants and animals, and gave each a unique name. Keeping a library of specimens allowed naturalists to further define and refine the identification of species. It was also Linnaeus who introduced the uniform, hierachical system for naming species that is used all over the world today. His system, now much more refined, is known as *Linnaean taxonomy*, and he wrote much of his work, and named the organisms he discovered, using Latin, or Latinised Greek. It is often said *Deus craevit, Linnaeus disposuit* – 'God created; Linnaeus put it in order.'

One of the most important aspects of the Linnaean system is how each living organism, once it has been classified, is named, using *binomial nomenclature*, which refers to a distinct species by the two lowest ranks to which it belongs, its genus and species.

TABLE 3.1. Linnaean Classification of Two Bee Species

Species Name	*Bombus terrestris*	*Andrena fulva*
Domain	Eukarya	
Kingdom	Animalia (animals)	
Phylum	Euarthropoda (anthropods)	
Class	Insecta (true insects)	
Order	Hymenoptera (bees, wasps, ants, and sawflies)	
Family	Apidae (social and some solitary bees)	Andrenidae (solitary ground-nesting bees)
Genus	*Bombus* (bumblebees)	*Andrena* (mining bees)
Species	*terrestris* ('of the earth')	*fulva* ('reddish-yellow' or 'tawny')
Vernacular Name	Buff-tailed bumblebee	Tawny Mining bee

The *species name*, consisting of the full double-barrelled genus and species, or *binomial*, is always unique. The second part of the name, called the *species epithet*, may be used to identify species in many different genera, however. In fact, it is not uncommon for two very different species to share the same species epithet. The Tawny Mining bee, *Andrena fulva*, shares its species epithet with the tawny spider orchid, *Caladenia fulva*. They have nothing in common but their colour, which has nothing to do with how closely they are related. If two species share the same genus, however – such as the bees *A. fulva* and *A. cineraria* – then we know they are therefore closely related. *A. cineraria* is the Ashy Mining bee, a mining bee just like *A. fulva*.

It was not long after I got my head around why binomials work so well, and started using them more confidently myself, that my friend Stuart Roberts told me a story that made me chuckle. 'I used to tell the children I taught, that the Linnaen system was much more helpful than ours' said Stuart 'because it was patently untrue when football fans would sing, "There's only one David Beckham," when there are likely to be quite a few.

However, if they were ever minded to chant, "There's only one *Bombus magnus*," they'd be quite correct!'

Even though my grasp of biological taxonomy and binomial nomenclature is still rudimentary, I have gained a multitude of interesting insights about the plants and animals we share this planet with. It has, for instance, helped me understand why Labradors and Poodles can breed with each other, but robins and blackbirds cannot. To anyone with a scientific background, this would be blindingly obvious, but to me it hadn't been. I now know that whilst animals belonging to the same species can interbreed (all dogs are the same species), it is unusual for animals of different species to do so and have their offspring survive, or be fertile. Exceptions such as horses breeding with zebras, or tigers with lions, are rare and usually occur only in captivity.

It has also been eye-opening to discover which of our common garden bird species are related, and how closely they are related, simply by looking up their scientific names. Take the finch family, Fringillidae. It is fairly easy to pick out birds that belong to this family – including bullfinch, greenfinch, chaffinch, brambling, and siskin – because they all have distinctive stout, triangular bills. But although these birds belong to the same family, they do not all belong to the same genus. Until I looked up their full species names, I would never have guessed that a chaffinch (*Fringilla coelebs*) is more closely related to a brambling (*F. montifringilla*) than to a bullfinch (*Pyrrhula pyrrhula*).

My newfound knowledge of biological taxonomy allowed me to identify bees much more methodically – and swiftly. Instead of clicking on link after link in the BWARS species photo gallery in search of lookalikes, I downloaded their 'Bees in Britain: An Introductory Text to British Bees' and started studying. I diligently began to work my way through it, beginning with *A* for *Andrenidae*. However, with no mentor to guide me, I got bogged down in all the minutiae about each species and never quite made it to *M* for *Melittidae*. I decided that rather than overwhelm myself completely with information on all six bee families, I would concentrate on those genera with which I was already somewhat familiar: *Bombus* (bumblebees); *Andrena*, *Colletes*, and *Melitta* (solitary mining bees); *Megachile* (solitary leaf-cutter bees); and *Osmia* (solitary mason bees). This proved far more manageable.

I read about these species' sizes, ranges, and foraging preferences, their nesting habits, their physical characteristics, and much more. Then I went out 'in the field' – which often meant getting out into my garden – to observe bees and see if I could match them up with what I could remember. But there is still only so much information a brain can take on board. I didn't trust that I was getting my identifications right. I needed to check in with someone who had more expertise. If I had started heading in the wrong direction, they would put me back on track.

It was around this time that I joined Twitter, hoping to connect with other bee enthusiasts. I cannot imagine how my knowledge of bees could have developed as much as it has without the help I received, and continue to receive, from the group of expert and amateur naturalists I have met through this platform.

My earliest forays consisted mostly of me posting the odd photograph and plaintively typing, 'Please can anyone help me identify this bee?' I would then wait rather naively for a kindly entomologist, or perhaps the universe, to answer. Unfortunately, I had no real concept of how Twitter worked in those days, so, apart from my friend Jon, who helped me set up my account, and my children who followed me because I am their mum, no one else was actually seeing my tweets.

Once I realised Twitter was all about 'following' and 'being followed', things improved rapidly. I learned to search for people who regularly used the word 'bee' and 'followed' them. And to my delight, some of them followed me back. My circle of 'wildlife' friends on Twitter widened, and by 2011 our little group included a few experts who generously helped us out whenever we got stuck. Each of them encouraged and corrected us, but none so kindly and patiently as Ian Beavis, an entomologist and museum curator from Tunbridge Wells.

My exchanges on Twitter dramatically changed my approach to learning about our native bee species. I had been trying to learn by heart the names and characteristics of as many bee species as I could hold in my goldfish-like memory. Each time I came across a species I didn't recognise, I would photograph it from as many different angles as possible, upload the photos to my computer, and scour the internet and books, looking for a match.

Some bee species look exactly like others to the human eye and can be properly identified only by looking at them under a microscope; I was not personally prepared to kill the bees I found in order to identify them. I am an insect enthusiast, not an entomologist, so I would be killing these creatures purely out of idle curiosity, unlike the scientists who need to study preserved specimens to understand changes in ecology. What else could I do then? When all else failed – and more often than not, it did – I would upload my photos to Twitter and ask for help. I still do.

I have learned a great deal as a result, including a bit of humility. I vividly remember one day when I stopped to check out a patch of dandelions on the bank behind my house. I always stopped there. It was almost midday and the dandelions were wide open and showing off that bright golden yellow hue that makes people like me want to jump for joy and (most) gardeners and councils reach for the mowers and herbicides. The dandelions were covered in bees and hoverflies. They always are.

I noticed a couple of Red-tails and maybe half a dozen Common Carder bumblebees. There were some tiny solitary bees, too – *Lasioglossum*, I thought – and a larger female mining bee whose legs and sides were completely caked in pollen. She had a shiny black abdomen and a gloriously yellow thorax. I had seen most of these other bees before, but who was *that*? I ran inside to get my camera.

I felt sure I would be able to get back to the dandelion patch before the bee flew off. It's usually quite easy to take photographs of bees on these plants, because the flowers are *inflorescent*, which means they are composed of many tiny, individual flowers presenting as a single flower head. So instead of flying off to another flower as soon as they have supped the nectar, the bee usually just crawls around the dandelion, dipping its tongue into each and every individual floret as it goes. It's good evolutionary economics, as it saves both time and energy. I managed to take at least a dozen shots before my exciting new bee flew off.

Back inside, I uploaded and cropped the photographs. I could tell from the shiny abdomen and the way the pollen had been collected on the bristly hairs of her back legs that she was a mining bee of some sort, so I went directly to the *Andrena* section of the BWARS guide, looking for her. Three times I went through the images, but still I could not find a match. Had the hairs on her thorax been ginger or orange, there would have been a number of candidates, but they were bright yellow. I was stumped. I started to wonder if perhaps I'd seen a bee new to Britain and Ireland. Emboldened,

I posted my best photos on Twitter with the caption, 'Anyone know who this is? Could it be a new species?!'

None of my trusted entomologist acquaintances was online, but a few of my wildlife friends became very excited for me. By late evening I was checking Twitter every ten minutes or so, but still I'd had no response from any of the experts. I went to bed dreaming of giant bees with shiny black abdomens and bright yellow thoracic hairs.

As soon as I woke, I switched on my laptop and whizzed through the Twitter notifications. There was a reply from a trusted source! 'Brigit,' it read, 'this is an Ashy Mining bee covered in pollen.' *Oh.* To say I was embarrassed would be an understatement. How could I have missed something so obvious?

But it was a marvellous lesson. I had been so intoxicated by the thought of discovering a new bee that I did not pay any attention to the clues around me, such as the time of year, the surrounding habitat, and the other flowering plants in the vicinity. It is much easier to narrow down your search if you first ask the question, *Which bees might I expect to find at this time and in this location?* As Ashy Mining bees enjoy dandelions, nest in habitat like this, boast shiny black abdomens, and fly at this time of year, they should have been one of the first bees I looked up, given the facts.

You may have gathered by now that I enjoy learning. I was the child in class who hung on the teacher's every word to make sure I didn't miss a thing. I used to read our set textbooks from cover to cover and memorise their contents word for word, much as you would if you were learning lines for a play. I especially enjoyed it when teachers deviated from our textbooks, though, and I always had a notebook at hand to scribble down any extra titbits that were thrown our way, reasoning that these snippets of information might make the difference between a minus and a plus in my marks.

I remember at the age of ten or eleven getting up as early as 5am to steal a few extra hours revising in the boarding house bathroom during the weeks leading up to summer exams. I desperately wanted to do well, partly because I hoped to please my teachers and my mum, but mostly because I actually enjoyed the process by which you start off knowing absolutely nothing about a given subject and end up knowing *something.* And it paid off. I did extremely well in my exams. In essence, though, I was simply learning by rote, regurgitating what was written in the textbooks at the end of each school year. I don't remember ever questioning the material I was taught, or wondering what it all meant, let alone thinking about how I might apply it to real-life situations. Looking back this strikes me as odd, because outside the classroom, I used to question everything, driving my mother to distraction with all my whys and what-ifs.

As things turned out, despite all my hard work, I did not embark on any form of higher education, leaving school halfway through A levels. Perhaps that's why my rather Victorian view of what it meant to be 'educated' stuck with me until another exchange on Twitter a few years ago.

It was February and the weather was so unusually warm and glorious that I decided to walk over the hills instead of following my usual commute, down the West Malvern Road, to the organic café, Greenlink, in Great Malvern, where I was then working the afternoon shift. Before heading out, I tucked my camera in my rucksack, on the off chance that I might spot an early queen bumblebee on the wing.

There were no bumblebees on the hills that day, but as I walked down through town, I noticed something buzzing around some viburnum flowering by the side of the road. It was a male solitary bee of some kind, which excited me enormously because of how early in the year it was. The bee was not very active, so I was able to take a few snapshots of it before hurrying on to work.

When I got home that evening I did my usual thing: I uploaded the photos to my computer, cropped them, and searched for a lookalike species in the BWARS guide. I guessed it might be an *Andrena* species, but beyond this I had no idea. Scanning photo after photo, I finally found a close match, *A. nitida*. But BWARS noted that this bee did not emerge until much later in the spring. Tentatively, I posted my photos on Twitter and asked my bee followers for confirmation. My twitter friend Jane Adams, who knows far more than I do about solitary bees, thought it was. Still, we wanted to wait for Ian Beavis to come along to verify our identification.

In due course, he did. But Ian also surprised us: He told us that this particular *A. nitida* had been 'stylopised'.

Stylopised? What in the world did that mean? Ian drew our attention to a flattened, grublike protrusion between two of the bee's tail-end abdominal segments. At first I couldn't see it, so Jane zoomed into my image further and circled it in red for me. 'That,' Ian explained, 'is an adult female *Stylops*.' *Yuck*, I thought.

Later that evening Ian posted a drawing of an adult male *Stylops* that he'd found in a nineteenth-century book. *Stylops* larvae hitch a ride on adult female bees that are collecting pollen for a brood nest. The *Stylops* larvae burrow into the bee larvae and live inside them as parasites. Once they are fully formed, the male *Stylops*, with big, bulging eyes and squat forewings shaped like clubs, wriggle their way out of the body of the adult host bee in order to mate with a female. The female adults are wingless and dependent on their host's body, their heads peeking out from the abdomen so that the males can find them. Her eggs hatch inside the bee host, too, and then are carried to a flower, where the larvae are left to await a new host home.

Nature, as well as being beautiful, is also ingenious, and – from an anthropomorphic perspective – it can sometimes be very brutal.

Having read all that I could stomach, I went to sleep. I woke in the middle of the night from a nightmare in which I, too, had been stylopised. It was gruesome. I turned the light on, made myself a hot drink, and sat in bed wondering if perhaps the creators of the 1979 film *Alien* had based their terrifying monsters on the life cycle of this creature. That's when it came, my eureka moment.

How, I wondered, *did Ian notice that almost invisible and minuscule brown bulge between the abdominal segments on my blurry photograph of a flying bee?* I had struggled to see the bulge even after he told me it was there.

The following day I asked him. Ian explained that this particular species of bee doesn't usually fly until March or April, but bees that have been stylopised usually emerge earlier, whatever the weather. This is because *Stylops* larvae eat their fat reserves, leaving the bee starving for food. Given the time of year, he had immediately searched for the telltale bulge of the female stylops head between the abdominal segments, because he knew this would account for its early emergence. *Brilliant*, I thought. *Ian is a detective!* Of course, Ian would not think of himself as a detective. He is an entomologist, and entomologists, like all scientists, are always on the lookout for evidence to support or reject a hypothesis.

In a single moment, the horribly blinkered Victorian method of learning by rote that I had been taught in the 1960s, and continued to rely upon since, was evicted, banished forever to the place where tools and habits that no longer serve us get banished. In its place, I welcomed a new and wonderous understanding of what it means to educate and to be educated.

Having missed out on further education, I had rarely been asked to cultivate a larger curiosity about my subjects, or to apply the facts I memorised to the world beyond the pages of my textbooks or exam papers. Henceforth, I resolve to embrace lateral, creative, and inventive thinking, as well as an inquisitiveness about, and wonder and awe of, whatever subject takes my fancy on a given day.

The word 'study' had taken on a new meaning for me.

CHAPTER 4

The Boys Are Back in Town

The boys are back in town! I have just clocked them in Diana's vegetable garden, a marauding gang of moustachioed males fuelling up on warm, sweet nectar from the flowering chives before heading off on their scent-marking patrols. They are male bumblebees – to be precise, *Bombus pratorum*, commonly known as Early bumblebees.

Early by name, and early by nature, the queens of this species steal weeks, in some cases months, on other bumblebee species in the race to establish nests at the very beginning of the season. There are other species – Buff-tailed bumblebees, for instance – that may emerge from hibernation earlier, but it is almost always the Early bumblebee nests that produce the first males and queens. Their nests, as you might remember, are shorter-lived than most bumblebees', averaging just fourteen weeks compared with around eighteen weeks for other species.

Not only is this the smallest bumblebee in Britain and Ireland, but its nests are small, too, containing sometimes as few as fifty individuals, even at their peak.

I have been looking out for male bumblebees for a few weeks now, so I am overjoyed that they have finally arrived. Why their appearance fills me with quite so much joy, I cannot say. It's not as though I have been starved of bees for the last few months; it is now early May, and my computer already contains hundreds of photos of bumblebees and solitary bees I have seen so far this spring. Whatever the reason, seeing these little male bumblebees on the chives fills my heart almost to bursting. It's a good job no one can see me; it would be hard to explain why I am dancing, on my own, in Diana's garden. My bee friends on Twitter would understand. We watch each other's timelines with bated breath at this time of year, awaiting images of the first male bumblebees. Not in a *competitive* way, you understand – though I must admit I enjoy it tremendously when mine are amongst the first male bumblebee sightings of the season.

Could the knowledge that new queens will soon appear, and that matings will soon take place, be the reason I am so elated to see these males amongst the chives? The appearance of *reproductives* – queens and males – heralds the beginning of the final stage in a colony's life cycle. They are the surest sign that the colony has been successful.

I am acutely aware of the myriad challenges bumblebees face before they reach this final stage. It is a small miracle that they ever do. Starting from the moment new queens leave their natal nest, there are hurdles to overcome. First they must mate and locate a suitable hibernation site. They must then survive the winter, find sufficient food when they wake up in the spring, establish secure nests, rear armies of workers (all females), and then, only then, produce daughter queens and males that might mate and begin the cycle anew. At every step, the colony might not make it. So any sign that a nest has been successful is worthy of major celebration. And at the very least, a little dance.

One of the Early 'bumbleboys' has come to rest on the edge of the raised bed just beneath the chives. I have been watching him clean his antennae, but he is quiet now, taking a well-earned rest from his busy cruising circuit as he basks in the early summer sunshine. He is not in the slightest bothered by my presence, so I gently put my hand down beside him and invite him to climb onto my palm, to get a closer look at him. He crawls aboard and his feet tickle my skin. I know he won't sting me. Male bees do not possess stinging apparatus.

My Early bumblebee male is ridiculously cute: small, rotund, and kind of scruffy compared with the females of his kind. Like the Early queens, and workers, he sports a tail that is dull orange in colour, with a black band across the lower part of his abdomen and a thick yellow band across the top of it. In the female worker bees, this yellow band is often missing, or so dull that you can barely see it, but as my bee is a male, his yellow bands look brighter, fresher, and more lemony. His thorax is black, and then there is another thick, bright yellow collar, like a miniature lion's mane, just behind his head. Were I a female bumblebee, I believe I might already have fallen a little in love, but as he turns to face me, the female bumblebee in me goes weak at the knees as I catch sight of his yellow, Chaplinesque moustache. Yes, this delightfully dapper bumbleboy sports a moustache.

Bumblebee identification is a little trickier than you might imagine. Although there are now only twenty-four different species of bumblebee in Britain and Ireland, there are still a great many variations, not only between species but also within species. Just when you think you have got to grips with the more common species that visit your garden, you realise that the workers are not necessarily cut-down versions of their queen; that the males are sometimes entirely different in appearance to the workers *and* the queens; and that there are different forms, melanisms, aberrations, and subspecies of each. As if these were not confusing enough, the bees' colour begins to fade with each week of foraging in the bright summer sunshine.

Male bumblebees, thankfully, have a number of distinguishing features that make them fairly easy to identify as males. Most obvious is their facial hair, especially in those species, like the Early bumblebee, that sport those handsome little moustaches. But there are also other physical differences between males and females (what biologists call *sexual dimorphism*). Queen and worker bumblebees have beautiful shiny, flat, and smooth pollen baskets, called *corbiculae*, fringed with long hairs, on the outside surface of each hind leg which often – but not always – contain pollen. Male bumblebees, on the other hand, never collect pollen, so have no need of a pollen basket. Their legs are noticeably rounder and hairier, and matt rather than shiny. As well, females have v-shaped back ends (they come to a narrow point where the sting comes out) whilst males have blunt ends with a truncated tip. Finally, though you'd need a magnifying glass to check this feature, male bumblebees each have an extra segment on their antennae, making these ever so slightly longer than those of the females.

Apart from the physical differences between the sexes, there are also differences in behaviour, the most significant being that males do absolutely nothing to contribute to the well-being of the colony, their sole purpose in life being to mate. Perhaps this is why, once they are ready to leave the warmth and comfort of the nest, they are rarely allowed back in again. Whilst the workers spend their entire lives 'working' – foraging for food, feeding the grubs, cleaning and guarding the nest – males spend their lives patrolling and scent-marking rendezvous sites, refuelling with nectar, patrolling again, and hoping to find a queen with whom to mate.

As they have no nest to go home to, male bumblebees can often be found late at night, or early in the morning, sleeping inside and underneath flowers. It might sound like these boys live the life of Riley, but spare a thought, when the weather is cold and damp, for how tough it must be to sleep out on the tiles whilst your sisters go home to roost inside a warm, cosy nest. Worse still, very few male bumblebees – only one in seven – ever actually achieve a mating.

Interestingly, male honeybees, or 'drones', whose sole purpose in life is to mate with new honeybee queens, are permitted to come and go from their hives at will. Unlike male bumblebees, honeybee drones are incapable of collecting food for themselves, so are provided for and fed by the female workers. Again, theirs sounds like an easy life, but the downside to being a honeybee drone is that, for those lucky few that get to mate with a queen – an encounter which occurs, mid-air, during her nuptial flight – the mating lasts for only a few seconds. After this, his sexual organs and abdominal tissues are ripped violently from his body as part of the process, a gruesome death if ever there was one. There are no consolation prizes for the losers, either. Once the mating season is over, and the male population has served its purpose, those that remain are pushed out of the hive by the female workers. Unable to fend for themselves, they soon die of cold or starvation.

And what about solitary bees? Well, with one or two exceptions, male solitary bees usually emerge from their nest cells a week or so before females of their species, and spend the next few weeks foraging, and patrolling nesting sites or popular floral resources, in search of females. As you can imagine, with some twenty thousand different species of solitary bee in the world (some less 'solitary' than others), courting rituals and mating behaviour vary enormously, and scientists still have much to discover about both.

Broadly speaking, where there are dense nesting aggregations with many females emerging at around the same time from their natal nests, such as happens with Ivy bees (*Colletes hederae*), males maximise their chances of reproductive success by hanging around the nest entrance for the females to emerge – what is called *nest-guarding*. Then, as each female emerges, she is pounced upon by multiple males forming fiercely competitive 'mating balls'. During the ensuing scramble, one of the males will succeed in securing the female and mating with her. Although I have watched this behaviour many times with different ground-nesting bee species, I have not yet been able to determine whether or not the same males go on to mate with other females.

With solitary bee species whose nests are more widely dispersed, such as Wool Carder bees, there is no mass emergence, and nest-guarding does not make any evolutionary sense. Males of such species are far more likely to hang around the females' preferred foraging sources, in the hope that they might strike lucky there.

Generally speaking, after all the female solitary bees in the area have mated, there is no reason for the males to hang around the nest sites or food sources any longer. They live out the rest of their short lives drinking nectar by day and sleeping by night. They mostly shelter on their own, inside wall cavities or flowers, but they sometimes roost together in groups, gripping onto plant stems or seed heads with their mouthparts until the morning sun warms them up sufficiently that they can fly again.

No male bee of any species survives into the winter. Given the givens of mating, if I were a male bee, I think on balance I would rather be a solitary bee or a bumblebee, than a honeybee. Maybe a cute Early bumblebee, like my friend in Diana's garden.

For a few weeks, I go back regularly, eagerly watching the Early bumblebee males on the chives, foraging alongside the female workers, who are also very partial to this flower. The males are not even slightly interested in these sterile workers. I keep my eyes peeled for emerging queens in the hope that they, too, might visit the chives, and that I might catch them mating. My watch at the chives might sound a little naive to those who have studied bumblebee mating behaviour, but I am not so green as I am cabbage-looking: I have seen this species mate on flowers this height from the ground before. Added to this, the males I am watching today seem to be spending a lot of time flying back and forth from different clumps of chives with great purpose, and without stopping to refuel. I am feeling quite optimistic that the vegetable patch in Diana's garden might prove to be a rendezvous area.

Different bumblebee species are known to adopt different mating strategies. Like solitary Ivy bees, some of them nest-guard, lurking outside specific nests, hoping to mate with new queens as they emerge. Others employ *perching*, where they wait, perched upon a prominent object, ready to pounce on a passing female (or anything else that flies past for that matter). The rest adopt the 'patrolling' strategy, which involves spending the day flying up and down a scent-route, or around a rendezvous area, in search of virgin queens of their kind. With the exception of the Tree bumblebee (*B. hypnorum*), all male bumblebees in Britain and Ireland patrol.

The choice of rendezvous area differs from species to species. Red-tailed, Buff-tailed, and White-tailed bumblebees, for instance, are known to mate high up in the treetops, whilst Early bumblebees usually mate within a metre or so of the ground – about the height of Diana's chives. Dave Goulson and his colleagues at the University of Sussex have observed the males of some species, including Common Carders (*B. pascuorum*) and Bilberry bumblebees (*B. monticola*), congregating on hilltops, a type of patrolling known as *hilltopping*. Regardless of their choice of rendezvous spot, males of all patrolling species spend an inordinate amount of time and energy cruising their circuits and scent-marking them with queen-attracting pheromones.

Whether they are patrolling tree lines, hedgerows, hilltops, or patches of chives, male bumblebees become so intent on their purpose in life that they often forget to stop for breakfast, elevenses, lunch, afternoon tea, and dinner. Who can blame them? With seven males competing for every queen, taking time to sup on some nectar could be the difference between fulfilling their purpose and dying without offspring. If you happen upon a male bumblebee exhausted by his quest to mate (or any other exhausted bee, for that matter), you can help revive him by offering him a few drops of a 50/50 sugar-water solution. Counterintuitively, a solution made from white sugar and water is safer to give a bee than honey is. Honey, though always safe for humans, might occasionally contain viruses or fungal spores that can spread disease from one bee species to another.

I wish I had known about the patrolling strategies used by male bumblebees earlier. Back when I lived on the Malvern Hills, for a number of consecutive years I noticed male Red-tailed bumblebees congregating in and around a copse of rowans halfway up the hill just above West Malvern Road. Every single year they chose the same clump of trees. It fascinates me now to realise that this must have been a rendezvous point for them, and that the males' scent-marking pheromones were species specific, and strong enough so that newly emerged Red-tailed queens in the neighbourhood were able to detect them and find their way there.

I wonder how far queens travel to find the males, and how many rendezvous points are established for each species in any given area. I have searched for answers to these questions, but compared with other aspects of bumblebee ecology, I have found it more difficult to find information about their courtship and mating behaviour. One thing in particular that I was hoping to discover more about if and when it occurs, is the extent of nest inbreeding. Suffice to say, when it comes to bee mating, 'It's complicated.'

I get the feeling that, until recently, scientists have perhaps given male bees less attention than the females. The attempts I have made to discover more about male bumblebees have often led me back to Laura Brodie's excellent website, bumblebee.org. Here I learn that the pheromones given off by some male bumblebee species can actually be detected by humans. Had I known this whilst I was watching the male Red-tailed bumblebees on the Malvern Hills, I would have made a point of trying to pick up their scent, which in this case smells of citronella.

In any case, I am now on the watch for any and all signs of male bumblebees patrolling. Which is why I am so interested in the Early bumblebees gathering around the clump of chives in Diana's vegetable plot.

It is probably because I am paying so much attention to the chives that I notice two, three, then four days in a row, a growing number of Early bumblebees on the ground beneath them, all dead. I examine the carcasses but I can see no marks or clues as to how the bees have died. They are brightly coloured, with wings in pristine condition. This tells me that they have not died of old age; older bees tend to be faded in colour and have raggedy wings. Nor has it been cold, so they have not frozen to death. It is a mystery.

Simultaneously, I notice that fewer bees, of any species, are foraging on the chives. I can't help wondering if the Early bumblebee deaths might have been caused by the flowers themselves. If the chives were in any other garden, I would suspect pesticides, but Rob takes care of this garden, so death by pesticide is inconceivable.

By the fourth day, hardly any bees are visiting this clump of chives, though other nearby clumps are as busy as ever. On this day, though, as I walk through the vegetable patch towards the chives, I spy an Early bumblebee worker on the 'ghost clump'. I draw closer and see that she is not moving. I watch for a while longer before gently touching her with my finger. She falls to the ground, dead. But as she falls, I notice a movement on the flower from which she fell. It is a spider. A white spider. That looks like a crab. Oh, my goodness, I know what is going on here!

The creature is a crab spider (*Misumena vatia*) – a fearsome predator, sinister lurker, ambusher, and master of disguise. Poor Early bumblebees, they didn't stand a chance against this awesome foe. It hunts by stealth, sitting motionless in a flower, and (in the case of mature females) often taking on the colour of the petals like a chameleon, until an unsuspecting insect lands on the flower head to feed on the nectar. That is when the spider strikes, pouncing on its victim, sinking its fangs into the insect's back, and injecting a paralysing venom. Size is no deterrent to this spider, who will attack and eat all manner of insects, including bumblebees, flies, and moths at least twice its size.

Crab spiders do not have teeth, so after they have successfully paralysed their prey, they liquefy the insides in order to drink them, leaving the exoskeleton intact. This explains why the dead bees I found on the ground looked so perfect and unmarked.

I consider moving the spider to another part of the garden to protect the Early bumblebees that feed on these chives, but I decide to leave her where she is. It is clear that the bees have already become wary of landing on this particular clump. Perhaps the dying bees gave off some kind of pheromone to warn others not to land there. I don't know, but in the end I think it would be wrong to interfere. I abandon my daily watch for newly emerged queens. Despite my decision to leave nature to do what nature does, I prefer not to witness the crab spider in action. I am not even sure that the males are patrolling here any more. I reason that if I am meant see a mating this spring, I will.

Later that very afternoon, the universe answers when I find a pair of Early bumblebees mating on the phacelia on our allotment. I have no way of knowing how long this pair have been mating, but from the moment I first spot them, they remain locked together for at least another twenty minutes, during which time the queen crawls from one flower head to another, feeding as she goes. This does not require much effort on the queen's part, as the male is quite small in size compared with her. With some bumblebee species, especially those that produce the most enormous queens, the male can be just a third of her size. The size difference between queen and male Early bumblebees is not quite so pronounced.

With bumblebees, the transfer of sperm takes place within the first few minutes of mating, although the male remains attached to the female for a long time afterwards. Recent research suggests that following insemination, the male inserts a 'plug' that, when it hardens, blocks the queen's genital opening for at least the next few days, increasing this male's chance of passing his genes on to the next generation. Given the ratio of males to queens, it is highly unlikely this male bumblebee will ever get a chance to mate again.

Because it is still so early in the season, this newly mated Early bumblebee queen will almost certainly form a new colony of her own, but the daughter queens that emerge from her colony in a few months' time will, after they have mated, probably go straight into hibernation for the winter. Early bumblebees are an exception rather than the rule when it comes to having two (or occasionally three) complete life cycles in one year. Most bumblebee species in Britain and Ireland complete only one mating cycle each year.

I am looking forward to seeing other male bumblebees emerge as the weather warms. I find it difficult to tell the difference between the sexes with some species – Common Carder and Tree bumblebee males look, to me, just like the workers of their species. But I can spot Red-tailed and

White-tailed males from a mile off. As with the Early bumblebees, their appearance differs considerably from their kindred queens and workers.

At some stage I would love to improve my skills in identifying male bumblebees, but for now I am happy to look out for the hairy legs, bright yellow bands, and mini moustaches that tell me the boys are back in town.

CHAPTER 5

Bees Behaving Badly

In the very top left-hand corner of our allotment, on the border between our plot and the cottage gardens that back on to it, lies a large patch of Russian comfrey. No self-respecting vegetable plot should be without at least a small patch of comfrey, I say, whether Russian, native, or hybrid. Enricher of soils, fertiliser of Rob's tomatoes, component of healing balms and salves, and provider of pollen and nectar for numerous species of bumblebees and other insects, this plant is right at the top of our list of essential perennials.

As a soil enricher, comfrey provides significantly higher quantities of potassium, potash, and nitrogen than other organic fertilisers, even out-performing most manures, composts, and liquid feeds for concentration of nutrients. In fact, so beneficial is this plant to organic growers, that in 1958, when the father of organic gardening, Lawrence Hills, founded the Henry Doubleday Research Association, now known as Garden Organic, he named the association after the nineteenth-century Quaker smallholder who reputedly introduced Russian comfrey (*Symphytum* × *uplandicum*) to Britain. I say 'reputedly' because, when you dig a little deeper, it appears the herb had already been introduced to England at the end of the previous century by

Joseph Busch, who was Catherine the Great's head gardener. Busch was a great fan of Russian comfrey, planting it as an ornamental in the gardens of St Petersburg Palace, but fell in love with it for its healing properties and sent roots back to his homeland. Apparently, the comfrey roots used by Henry Doubleday in his early experiments were taken from the gardens of St Petersburg Palace, too. Regardless of who deserves credit, I take my hat off to Joseph Busch, Henry Doubleday, and Lawrence Hills for the parts each of them played in introducing and promoting Russian comfrey far and wide.

Just below our patch of comfrey, behind our greenhouse, we have a large blue barrel with a tight-fitting lid in which Rob brews comfrey tea. 'Comfrey tea' sounds deliciously wholesome and nutritious, and so it is, but for plants, not humans.

There are a number of ways you can make this all-singing, all-dancing brew. You can follow a recipe, measuring exact amounts of leaves and water. Or you can simply do what Rob and I do, which is to chop the comfrey back every now and then, stuff the leaves into a big blue barrel, and top the barrel up with water. We wear gloves as we do this, because the stems are prickly and slightly irritating. We then leave this concoction to steep for three to four weeks, after which time it's ready to be used as a liquid feed around the roots of our tomatoes.

As we make our feed in a huge container, the resulting liquid is already adequately diluted, so we are able to use it as it is. If, however, you use more leaves and less water, you may end up with a thick dark liquid once the leaves have rotted down. Before pouring the feed around any roots, you will need to dilute it; otherwise, it will be too concentrated and do more harm than good. (It's worth noting that a 'tight-fitting lid' is an absolute must, because once the leaves start to rot, 'comfrey tea' smells seriously pongy.) If you don't have the time, space, or inclination to make a liquid feed with your comfrey, you can just use chopped-back leaves as a mulch around the bottoms of your plants, or add them to your compost, where they will act as an enriching compost 'activator'.

Comfrey has been cultivated for more than two thousand years as a healing herb. Its botanical name, *Symphytum*, is derived from the Greek *syphyto*, which means 'grow together', and its common name, 'comfrey,' comes from the Latin *confimare*, meaning 'to unite or heal'. Both the Romans and the Greeks held it in high regard, turning to it whenever they had need to stop heavy bleeding or heal wounds and broken bones. No wonder it is also known as knitbone and bruisewort.

As well as being a provider of nutrients for plants and a healer of human wounds, comfrey also happens to be a fabulous source of nectar and pollen for numerous insects. According to a survey conducted by the AgriLand project and funded by the UK Insect Pollinators Initiative, comfrey is one of the top ten plant species for nectar when measured in micrograms of sugar produced per flower per day.

Fortunately, for pollinators and gardeners alike, you can extend the flowering period for this plant by cutting it back. There are other plants, such as *Centaurea* and *Nepeta*, that can also be encouraged to produce a second flowering if you chop them back after the first flowering, but with comfrey, you can cut it back at least three or four times, and it just keeps on flowering, from early May until the first autumn frosts.

If you like the idea of this plant, but don't have room for a sprawling patch of three-foot-high Russian comfrey in your garden, you could try growing one of the dwarf varieties. Dwarf comfrey seems to flower a little earlier than the larger varieties, too, which is great for late spring bees. I have noticed Hairy-footed Flower bees (*Anthophora plumipes*) often switch to dwarf comfrey around the time the lungwort flowering season is coming to an end.

Whatever the variety, the blooms of this plant resemble little tubular bells. They hang in a cluster at each end of a multiforked stem, blooming one by one, in succession, as the tightly curled stem gradually unfurls and eventually straightens. Wild comfrey flowers are usually creamy yellow-white or pinkish purple, but this species cross-pollinates, and the Russian comfrey that grows on our allotment – a naturally occurring hybrid of two wild species – produces flowers in varying hues of pink, blue, purple, and red.

Amongst the first bees to visit this plant at the start of its flowering period in mid to late May are Hairy-footed Flower bees, Early bumblebees (*Bombus pratorum*) and Buff-tailed bumblebees (*B. terrestris*). Other species soon follow. Within weeks our comfrey patch reverberates with the gentle hum of pretty much every one of the seven bumblebee species we commonly find on our allotment, together with honeybees, hoverflies, and wasps.

I remember being surprised, and a bit perplexed, when I first realised exactly how many different species of pollinating insects are attracted to this plant. I was living in Malvern and had a patch of dwarf comfrey growing at the end of my garden path that was always covered in bees. I was then still preoccupied with trying to work out which bee was which, rather than what they might be doing, but whilst photographing them, I became aware that something wasn't quite adding up. I'd never seen such a variety of bee visitors on a single plant.

The comfrey's bell-shaped flowers are long and deep, with fused petals, which means the opening of the blossom is quite tight. It was obvious to me that the Hairy-footed Flower bees could reach inside the flower tube, or *corolla*, to collect their rich nectar rewards, because they had a long tongue. So, too, for the Common Carder bumblebees (*B. pascuorum*) and Garden bumblebees (*B. hortorum*), the latter boasting one of the longest tongues – up to fifteen millimetres long, when fully extended – of any bee in Britain and Ireland. But what about the huge Buff-tailed and White-tailed (*B. lucorum*) bumblebees that were also making a beeline for my comfrey? Both of these species have short tongues. They could not possibly access the nectar inside a flower with such a deep corolla. What, then, were they doing here?

Today, on our allotment in Shaftesbury, I am no longer surprised to see bees with such varying tongue lengths on plants with long corollas. This afternoon there are dozens of bumblebees buzzing around our comfrey, far more than I can count. I can see Buff-tail, Early, Common Carder, and Tree bumblebees. There are honeybees, Red Mason bees, and a few wasps, too. Of all these insects, only the Common Carder bumblebees, and some Early bumblebees, are accessing the nectar the way you would expect them to – that is, via the opening of the flower. The others are landing on the outside of the flowers and heading in completely the wrong direction, away from the opening and towards the base, where the flower joins the stem. But these insects have not lost their way; they know exactly where they are going and what they are doing. And they are behaving in a most impolite manner.

Ignoring the mutualistic relationship that has existed for millennia between most plants and pollinators, these other insects are flouting the rules, exploiting the generosity of this magnificent provider and plundering its most precious liquid treasure, the sugary nectar, without giving so much as a single grain of pollen in return. They say that there is no such thing as a free lunch, but whilst the Common Carders are conscientiously entering the blooms through their doorways, the rest, I am sorry to say, are breaking the law. In short, they are committing larceny.

Floral larceny, or *nectar robbing*, is a common phenomenon, with certain plants, such as comfrey, more likely to be targeted than others. Have you ever noticed tiny holes in the spurs of your columbine blooms, or in your delphinium, foxglove, honeysuckle, snapdragons, or runner beans? These holes are made by hungry bumblebees with short tongues. Unable to access the nectar rewards these flowers have on offer via nature's intended opening, they have found another way to relieve the plants of their bounty.

When planting a garden to cater for different pollinators, it is essential to choose a variety of flower shapes – some with bells and funnels, others with domes or cups, or flat – to match the range of abilities, and tongue lengths, which evolution has endowed different insects. But no matter how well we stock our gardens with different varieties, some bumblebees are just unable to resist breaking in to a blossom that they would never be able to reach the legitimate way, for a quick fix. These robber bees chew a hole through the tissue at the base of the flower, where the nectaries are situated, then insert their short tongue straight into the nectaries. By accessing the nectar in this way, they bypass the flower's stigma and anthers, so no pollination takes place.

Once the wall of the corolla has been breached, numerous species of insect – bumblebees, honeybees, wasps, and hoverflies alike – are free to take advantage of them for feeding. From what I can gather through speaking to gardeners, it is commonly assumed that each of these insects has, itself, chewed its way into the flower, but in fact the majority are opportunistic 'secondary' thieves. In other words, looters. The primary culprits – those guilty of making the hole in the first place – are none other than the Buff-tailed and White-tailed bumblebees, both notorious larcenists.

The eighteenth-century German naturalist Christian Konrad Sprengel, one of the founders of the science of pollination ecology, described these bees as 'committing an outrage against the flower'. Strong words, Herr Sprengel! But are these bees really behaving *badly*? And are the flowers they rob seriously compromised by their actions?

On the surface, it certainly appears to be a lose–lose situation from the plant's perspective. It takes a great deal of energy for a plant to produce nectar. And let us not forget that the sole purpose of nectar production is to reward insect pollinators; the plant itself has no use for this sugary treat. How then can a plant reward subsequent legitimate visitors, whose foraging might have resulted in pollination, if its nectar has been lost to larcenists? How could it ever attract enough legitimate visitors to bring about pollination, fertilisation, and seed set? Somehow it does, because otherwise the most frequently targeted plants, including those like comfrey where, in many cases, every last flower head has been robbed, would simply not survive. When I dig deeper, I learn some studies have shown that nectar robbing actually increases the rate of pollination of flowers, because legitimate flower visitors have to probe deeper into the flowers to access the nectar, making it more likely that they touch the stigma.

Since becoming aware that bees do not always enter a flower through its legitimate opening, I have also noticed another form of floral larceny, known as *base working*, wherein the bee steals the nectar by sneaking in between the corolla's petals. Honeybees seem to do this a lot. This form of larceny works only when the petals of the corolla are not fused, but it generally bypasses the reproductive structures of the plant, too.

I wonder why certain plants with long corollas are targeted, whilst others escape the burglars. Is there some kind of floral cue that marks some plants out as willing victims or easy targets? Do nectar robbers change the behaviour of legitimate pollinators? Can foraging bees distinguish between robbed and unrobbed flowers? I have whiled away many happy hours watching both primary and secondary robbers in our garden, on the allotment, and in the hedgerows and lanes of North Dorset and farther afield, trying to discover patterns. Alas, my observations have mostly given rise to more questions and very few answers. This is not surprising; from what I have read by scientists doing research in this area, there is much still to be learned.

The focus of my observations has been those plants I most regularly see being robbed: comfrey, red campion, and Rob's broad beans and runner beans. These plants bear absolutely no resemblance to one another, so it appears there is no 'typical' plant shape or structure that predisposes itself to being robbed. And given the abundance of comfrey and red campion near our home in Shaftesbury, it seems neither is suffering as a consequence of being a magnet for both primary and secondary larcenists.

My guess is that enough insects must still be accessing the nectar via the legitimate route, at least enough that the plants can take the hit from the nectar robbers. Also, as Rob pointed out one day whilst I was watching the robbers visiting the Russian comfey on our allotment, I have perhaps been focusing too much on the bees collecting *nectar* from its flowers. Of course, there are also bees that visit this herb's flowers specifically to harvest its pollen. Pollination is likely take place during their visits.

Plus, as far as comfrey is concerned, it can clone itself, or be cloned, via its roots. Technically, this means it does not need to be pollinated in order to reproduce. However, cloning is not a very sustainable form of reproduction in the long run, as the plant misses out on all the benefits of genetic diversity that come from cross-pollination. Still, because comfrey has the ability to clone itself, I am not able to tell how much impact the nectar robbers have on the pollination and fertilisation of this plant.

In the bees' defence, when they first start robbing the comfrey, early in the plant's flowering season, there is often a noticeable absence of flowering plants suitable for short-tongued insects in the vicinity. This scarcity has led me to wonder if the bees have been driven to thievery in much the same way as humans, or any other animal, might be when they are hungry. If so, thank goodness there is no Ancien Régime in nature, and bees, when they are hungry, can break the usual rules without fear of being beheaded.

Whatever drives them to a life of crime, after they have gained access to nectar the easy way, there's no going back, even though there may be other plants flowering that are better suited for their short tongues; the nectar rewards of those long corollas are just too seductive. So the robbers continue robbing, and other foragers watch and learn how to access the plant through the hole they have made.

In the case of red campion, which grows abundantly in the lanes and hedgerows where we live, I have watched garden bumblebees and various butterfly species nectaring through the entrance of the flower, and when I collect seeds in late summer and early autumn, I rarely come across any plants that have not set seed. These red campion are not being compromised by the larcenists, not in the least.

So far so good. However, Rob's broad beans and runner beans are an entirely different matter.

From what I have observed on our allotment, once the larcenists have made a hole, only the long-tongued Garden bumblebee (*B. hortorum*) continues to access the bean flowers via the legitimate opening. As local populations of Garden bumblebees are generally smaller than those of, say, Common Carder bumblebees, it is possible that Rob's beans are really feeling the effect of the nectar larcenists.

We face all sorts of challenges growing produce on the allotment, not only from nectar-robbing bees but also slugs, snails, sparrows, pigeons, mice, caterpillars, and goodness knows what else, so I can understand why commercial growers, whose incomes depend upon their crops being pollinated and harvestable, become frustrated when nectar robbers break into their plants' flowers. But if you have committed yourself to gardening organically, and with wildlife in mind, there are inventive ways to protect your flowers without resorting to the use of chemical pest control. To prevent larcenists from robbing all the nectar from Rob's beans, we have planted a large patch of phacelia nearby to try and draw them away.

Phacelia, like comfrey, is one of our must-grow allotment plants. In addition to being attractive to all manner of beneficial short- and medium-tongued insects, it is a nitrogen fixer, which means it enriches the soil, too. There is one problem, however: So attractive is phacelia to bees and other insects that you risk them ignoring not only the broad and runner beans but other plants that they would normally visit as well. For this reason, we plant just enough phacelia to draw the robbers away from the beans, but not enough that they are tempted to ignore the rest of our fruits and vegetables. We have no prescription for how much we plant; it varies from year to year, depending on how much space we have available. At the end of the day, you plant what you plant, and the bees and others insects will visit what they want.

Pollination ecology is one of the most interesting areas of study I have come across, and if I were a scientist, this is the field I would most want to work in. However, I have to confess that I am equally content to continue watching and learning from the bees themselves. And since my encounter with the stylopised *Andrena nitida*, I would go further and say I prefer *not* to have the answers to all my questions handed to me. Whilst I enjoy being able to extract gems of knowledge from scientific papers and bee guides, I also love the mystery – noticing the peculiar behaviour of a bee and looking for some clue that explains it. But more than that, I love the magic – those things that remain beyond the realms of science.

We can speculate, postulate, and theorise, but at the end of the day we are not bees. No matter how clever we humans are, there are aspects of bee behaviour we cannot possibly hope to understand.

CHAPTER 6

The Upside-Down Bird

'Age is just a number,' they say. My knees disagree. But at least they still work, and for this I am truly grateful. They complain a little when I walk on uneven ground or climb a hill, and protest bitterly when the path back down is steep and I have forgotten to bring a stick with me, but this is just a natural part of the ageing process. Despite these odd niggles I am still able to walk in the countryside for hours on end whenever I choose. I cannot bear the thought of not being able to walk, where I want, and when I want.

I am more aware of the precious gift of my mobility than I might have been had I not recently witnessed my mother's decline from being completely able-bodied and independent; to using a walking stick; then a mobility trolley and, after that, a wheelchair; through to being confined to her nursing home room and, ultimately, to her bed. Because of my mother's extraordinary accepting nature, she bore this journey without complaining. I doubt I could do the same.

My mother, Isabel, had many crosses to bear. In addition to her mobility problems, she suffered arthritis of the spine and pulmonary fibrosis, a debilitating lung disease that left her coughing incessantly and fighting for breath. She also struggled with dementia, and we never knew from one day to the next what state we would find her in. My mother was a highly intelligent woman, and was more troubled by the dementia than the rest of her ailments put together. Nothing scared her more than the thought that she might lose her mind. Thankfully for her, and for us, she never once stopped knowing who her family and friends were.

I was not a good nurse. I wanted to be, so much so that it tore me apart. I lie awake at night, still, riddled with guilt that instead of bringing my mother home and caring for her myself, I allowed others – admittedly people more qualified than me – to attend to her personal and medical needs, including, for the last few months of her life, her every need. I console myself only in the knowledge that I did the best I could. It may not have been someone else's best, but it was mine.

Outside of my work (and writing this book), I spent most of my waking hours fighting battles for her and trying to make her life easier. Foremost was finding ways to make sure she was eating. She had very particular tastes, my mum, and would eat very little of the food provided by the staff, so I experimented with different ingredients until I found things she would eat and enjoy. When she was no longer able to eat solid food, I made soups, smoothies, and lemon jelly; and when she could no longer swallow these, we gave her lemon tea and black coffee soaked into a little pink sponge attached to a stick like a lollypop. This, I learned, is how you help someone take fluids when she can no longer swallow.

I also dedicated myself to making my mum's nursing home room feel like a home from home. We filled it with family photographs as well as her favourite trinkets, pictures, ornaments, and lots of books. In her more active days, she had enjoyed flower arranging, so loved it when we brought in freshly picked flowers to brighten her room. She always asked their names, and whether Rob had grown them himself, which more often than not he had. Rob would chat to her about the flowers whilst I made her room look pretty.

She loved spending the time with Rob. 'He's such a kind man,' she would say when he left us alone. 'I'm so very glad you found each other.' I knew from her sister, my aunt Anne, that one of my mum's twilight joys was knowing that I had found love again, and that I would no longer be on my own. She used to worry terribly about me being alone.

I worried about her being alone, too. She received daily phone calls from Anne, who filled her in with news and gossip, and she enjoyed occasional visits from friends, but nothing lifted her spirits quite so much as seeing her family. I'm sure this must be the case for most people who live in care homes, or on their own. My brothers, Peter, David, and Patrick, and I would visit as often as we could. As two of my brothers lived overseas, in Australia and Dubai, their visits were all the more special.

Because I was unable to be with my mum around the clock, and sometimes couldn't visit for days on end, we arranged for beautiful, kind friends to call by to fill the gaps. They gave her reiki healing, massaged her feet, talked, listened, and held her hand. When my brothers visited, they helped her with *The Times* cryptic crossword, just as they had done when they lived at home. On Thursdays, she was visited by Deacon Michael, who lived in nearby Sedgehill. He brought her Communion, and prayed with her.

As her health deteriorated, my mother isolated herself more and more, refusing to leave her room for coffee or lunch unless someone she trusted offered to take her out, which, because of her frailness, became increasingly difficult. Also on Thursdays, Joanna, who had helped to care for my mum before she moved into the nursing home, came. Joanna helped her dress and put on her make-up before taking her into town in a wheelchair. Apart from my mother's obvious frailty, you would never have guessed how unwell she was if you saw her on Joanna's day – all warm and cosy in her bright emerald-green felt coat and the multicoloured Kaffe Fassett scarf she had knitted for herself sometime in the 1980s. And smiling, always smiling.

When these outings began to exhaust her, I searched for something that might distract and cheer her up on the days when none of us was able to visit. She was no longer interested in television or radio, or in reading the news. She even stopped picking up her beloved books. This saddened me more than anything, as she had been such an avid reader. There had to be something else, but what?

The idea to put up a bird feeding station came from something I had read about a project undertaken by Shropshire Wildlife Trust to place bird feeders outside the windows of elderly and disabled people, and have a network of volunteers fill them regularly. My mother and father (when he was still alive) used to enjoy feeding the birds in their cottage garden, piling their bird table – a square piece of wonky wood nailed on top of an old fence post

– with leftover crusts, bacon rinds, and the like. I suspect the rats got most of it, but birds came, too, and although I would never have called my parents 'birdwatchers', they certainly enjoyed watching those that visited them.

I sounded my mum out about the idea, telling her about all the birds that visited the feeders we had at home and suggesting that we put some up outside her window at the nursing home. She was all for it. So I ordered a very posh-looking feeding station, fitted out with four hanging feeders and two little dishes. I then stocked up on a variety of seeds, nuts, and fat balls from the local pet shop.

The feeding station arrived the very next day, and Rob lost no time putting it up on the patio outside her window. It was in a good position, with evergreen trees and a hedgerow directly opposite, which would provide perfect cover for the shyer garden birds. We rearranged the furniture in her room so that she could easily see the feeding station from her chair without straining her neck, then sat and waited for the birds to come. They didn't. It was a bit like watching a kettle that never boils, so Rob and I went home, leaving my mum propped forward in her chair with an extra cushion to give her a better view of her new feeding station – sorry, 'bird table'. My mum could not get her head around the terms 'feeder' and 'feeding station', and why should she? 'Bird table' it was, then.

As we left, I felt confident that the birds, when they came, would surely raise her spirits. And, oh boy, did they. Our phone was already ringing as we walked in through the front door. 'It's just me, love, I wanted to tell you we have our first visitor – a robin, I think. I'm watching him now, as we speak.' I listened as she described his behaviour. He was perching on the top of the bird table, not feeding yet, but checking it out, surveying his new domain.

My mum was clearly over the moon, not only because the robin had come but because she was able to name him. We said our goodbyes, but not half an hour later, the phone rang again. 'I can't tell you how much I'm enjoying this,' she told me. And again, ten minutes later, 'I think I must have scared him off.' She sounded so disappointed.

'I'm sure he'll come back,' said I, and sure enough, a few hours later, just as we were about to eat, the phone rang. 'He's back again, but he's chasing the other birds away,' she said. 'He's such a bossyboots!'

The bird table was clearly a resounding success – a mixed blessing for us, perhaps, because it generated a zillion more phone calls than we were already receiving, but an absolute joy and a whole new lease of life for my mother. We gave her the *RSPB Handbook of British Birds* for Christmas, and one of my brothers bought her a pair of binoculars. We spent hours together in her room, watching her new visitors, pointing out which they were in the RSPB book, and explaining, again and again, how to use the binoculars. To our knowledge she never once picked up the binoculars, or the book, when we were not there, since she kept telling us, 'I could do with some binoculars, and a book to help me identify all these birds.' 'It's a good job I've brought some in with me, then,' I'd say.

Apart from the robins, it turned out that my mother had forgotten the names of all the other garden birds, birds I had taken for granted she would remember. She kept on ringing me, describing her visitors as well as she could without her glasses (which she kept forgetting to put on) and without the binoculars (which she mixed up with her hearing aid charger and various other aids).

From what she said, she was getting regular visits from blue tits and great tits, as well as a small flock of long-tailed tits and some 'very noisy birds' that she would much rather had stayed away. 'They're pretty enough in their own way,' she conceded, 'with their speckled-y breasts and glossy green coats. But they're far too greedy for my liking. In fact, they're an absolute nuisance. Worse still they seem to be keeping the other birds away.' *Starlings*, we thought, and henceforth the robin was no longer the villain of the peace.

Then the goldfinches came, and oh, how these pretty little birds charmed and chattered their way into my mother's heart. She described them to us over and over again, delighting each time she saw them in the vibrancy of their colours: the flashes of yellow on their wings and their funny little berry-red faces. She thought they must be as beautiful as any bird she had ever seen. 'Like Fabergé jewels,' she said. We left the bird book open for her on the goldfinch page, so she could read more about them and remember the name. I don't think she ever looked at it. But she did ask me to buy a card with a goldfinch on the front so she could send it to her sister Anne, and proudly wrote in the card that this was the bird that most often visited her bird table.

There was one bird, though, that we had a job and a half identifying from my mum's descriptions. 'The upside-down bird,' she called it. Whilst other birds came to the table in their twos and threes, my mother only ever saw one of this particular bird, and apparently it didn't come very often. Because it was upside-down, as she kept insisting, she seemed unable to describe it. Nor could she tell us what it was feeding on, because she never did get to grips with the idea of there being different foods in different feeders. When we couldn't identify it for her, she got frustrated, and so did we. We began to think the upside-down bird might be a figment of her imagination.

Until, that is, the day it visited whilst we were visiting. Small, plump, with a chestnut pink breast and steely blue upper parts, it flew from the tree opposite my mum's window, straight down to the feeder with the nuts. Here, instead of perching upright on the side of the feeder as the other birds do, it clasped the wire with its feet, tail up and face down, and proceeded to pull nuts from the bottom of the feeder with its long slender beak. How we hadn't managed to work it out before is a mystery: a nuthatch. The upside-down bird was a nuthatch. It was so obvious when we finally saw it. Nuthatches are known for their knack of foraging tail up and head down. *Upside-down.*

I do not exaggerate when I say that my mother's avian visitors gave her more joy in the final few months of her life than I could ever have imagined or hoped. I wish with all my heart that Shropshire Wildlife Trust's scheme to place bird feeders outside the windows of elderly and disabled people could be rolled out everywhere. I had heard that the RSPB receive a great many unexpected bequests from people who have never been members. I used to wonder why people would choose *birds* over all the other worthy causes out there, but now I understood.

As my mother declined further, and spent more time asleep than awake, the daily phone calls with Anne, and Skype conversations with my brothers, became too challenging for her. Rob and I increased the frequency and duration of our visits, but the visits took their toll on me. I found it almost unbearable to see her, lying there like a little broken sparrow, so terribly frail and helpless, but somehow still smiling. I have never felt such deep sadness, and hope I never will again. 'You do know how much I love you?' I said every time she woke up, just in case I didn't get another chance. 'Of course I do, and I love you, too,' she mouthed. I wish I had not left it so late in my life to actually tell my mum I loved her, rather than assuming she knew. But we all have regrets, and if it weren't this, it would have been something else.

Towards the end, I asked her, during one of her more wakeful periods, whether she was frightened. She shook her head, smiled, and whispered, 'no,' with a gentleness and composure in her voice that I know was specially meant for me, to help me know it was all okay, that she knew what was coming, and that truly she was not afraid. And she wasn't afraid of death. There is no doubt in my mind that it was because of her faith that my mother was unafraid. I am not a religious person myself, but my mother's faith, which remained steadfast to the end, was an incredible thing to witness. It made me both grateful and envious. The strength and peace she drew from her beliefs radiated out from her diminutive human form to fill the entire room, so much so that it was almost palpable.

When she finally became too weak and fragile to leave her bed, we rearranged the furniture in my mother's room one last time. She could barely move her head from side to side then, so we moved her bed into the middle of the room, where she could look straight ahead and see the trees. We had earlier in the year sat down together to write her 'End of Life Plan'. Such plans are common for people who are terminally ill, or for elderly people who are living in care and nursing homes. They give you a chance, whilst you are still able, to tell people how you want to die, just in case you are unable to make your wishes known nearer the time. Whilst we were writing my mother's End of Life Plan, she told me that, for as long as she was still conscious, and if it were not too much trouble, she would like to see the trees.

The days were getting longer then, and the trees and hedgerows were almost in full leaf. We could no longer take my mum outside, so instead, we brought the outside into her room, filling it with cherry blossom, spring flowers, and greenery. We wrapped her up toasty warm, opened the room's patio doors wide, and wheeled her bed as close as we could to the doors so that she could see, and feel, the month of May in all its glory.

This was just three days before she died, though of course we didn't know that then. On the same day we opened the doors, a racing pigeon flew down from the tree opposite her room and landed on her little patio. He stationed himself there, feeding on the other birds' discarded seeds, and kept vigil with us till the end. He wandered inside the room a few times, too, which made my mum chuckle, and dropped a feather on her carpet. I picked it up and placed it in her hand. When my brother Peter next Skyped, I told him about the pigeon and asked my mum to show him her feather. She managed to hold it up, and when he asked her where it had come from, she replied, 'An angel, I think.'

Though we were easily able to fulfil her wish to see the trees, it saddened me more than I can say that I wasn't able to take Mum out to see the bluebells one last time before she died. She died on the 10 May, when the bluebells were at their best. But she missed them.

May was my mother's favourite month. 'Ne'er cast a clout till May is out,' she would say – 'clout' meaning warm winter clothing, but the word 'May', I believe, referring to hawthorn, or 'May' blossom, rather than to the month of May itself. She marvelled at the speed with which the trees and hedgerows greened up, changing almost daily from vague suggestions of buds about to burst to, suddenly and quite gloriously, being fully clothed. She loved this month and the changes it brings with it. So rapid are these changes that sometimes you can no longer see through a hedge that you swear was, only yesterday, still a brown-grey mass of tangled twigs and thorns.

Along with the greens come blankets of blue. We lived in Malvern for a year or two when I was around seven or eight years old, and on a Sunday afternoon after lunch, at my mother's request, my father used to pile us all into his old Volvo and drive us out to see, when they were at their most magnificent, the banks of bluebells that grew along the western slopes of the hills. It is no wonder that people come in droves from towns and cities to see the bluebells when they flower: They are intoxicating, intense, irresistible. By some curious twist of fate, I ended up coming back to the Malverns for a time whilst my own children were in primary school, and then again in my late forties. The second time, the front window of my little house in West Malvern looked directly across to the same bluebell banks we visited with my parents when I was a child. Deep down, I know it was the hills that kept pulling me back.

After the bluebells come the blossoms – cherry, crab apple, and hawthorn. The sight of their petals this year brings more memories, of a street in which we lived whose pavements were lined with cherry and crab apple trees. The flowers on these trees were pink, not a white in sight. From the way I heard my mother describing these blooms, I learned, at the tender age of five, that there are as many hues of pink in the world as there are letters in the alphabet, and that they have the most mouthwatering delicious names, from the palest blushed corals and pink lemonades, through to watermelons, raspberries, and rose, and on to the vibrant deep fuchsias that were my mum's favourites.

There was a lot of pink in our house. It was mostly contained in the dining room: curtains, candles, my mother's collection of vintage ruby glassware. Pink was the colour my mum most liked to wear, too. At least half the clothes in her wardrobe were some shade or hue. Even the shoes she wore every single day for the last few years of her life were pink – bright fuchsia pink, of course. And she wore them well. You could spot my mum a mile off, all four feet, eight inches of her.

But blossom in all its guises is short-lived. Blink and you miss it.

The early blossoms are already dropping, and the bluebells will soon wither and become engulfed in bracken. I need desperately to see them before they disappear. And so it is, in the weeks after my mother's death, that I seek solace in the hedgerows and lanes of North Dorset. I walk for miles and miles. Because I can. Up and down the narrow lanes I tramp, enveloping myself in spring.

At first, it is only the greens that I am able to take in. There are so many of them: dark, light, lime, grass, blue, yellow, soft, harsh, dappled, dusk, night-time. Although I went looking for bluebells and May blossom, I find I cannot see past all these greens. Perhaps I am afraid to notice the flowers, afraid of the joy they might impose upon my grief.

But slowly, gently, as my pace and pulse relax, the colours begin to seep in. First come cow parsley and hedge garlic, and in the hedgerow itself, the May blossom. Then greater stitchwort – constellations of pretty, white-petalled stars on long, leggy stalks, vying with red campion (which is not red at all, but pink) and bluebells on their last legs, but just about holding their own. I am relieved to see them and breathe a sigh of relief. I cannot let go of thoughts of my mum quite yet, but I can at least let the bluebells go. They will be back next year.

Finally, I acknowledge the yellows: buttercups, wood avens, nipplewort, lesser celandine, and yellow archangel. They are shiny bright and relentlessly, impossibly cheerful.

My spirits lifted, I begin to notice sounds and movements now, too. The familiar hum of a bumblebee, possibly a Common Carder, draws my attention to the understory. She is foraging on white dead-nettle. Of course she is, for this is what Common Carder bumblebees like best. I stoop to watch as she pushes her way deep inside the flower, until all I can see is her ginger-coloured behind. I must have watched hundreds of Common Carder bees pushing their way into hundreds of white dead-nettle flowers, and never tire of the sight. There is something strangely comforting about the familiar, something that tells us, 'All is well.' I wait, expecting the bee to visit more flowers in this little patch. But I must have happened upon her just as she was ready to head home, for suddenly she flies, straight up and over the hedge, without hesitation. She clearly knows exactly where she is going.

I stand back up again and look around. The sun is low in the sky and I have wandered a little farther than I intended. This particular lane leads out of Shaftesbury, south towards French Mill and Melbury. It has become extremely narrow now, banked up steeply on both sides with dark green growth that blocks out the sunlight. A tunnel. But it doesn't last long and soon I am out and back in open countryside.

The hedgerows on the far side of the tunnel are wild and unkempt, the trees given free rein to stretch their limbs, as much as a tree can when rooted in a hedgerow. A movement on the trunk of one of the trees just ahead catches my eye. I think at first it might be a squirrel scampering down the trunk, but then I spot the chestnut-orange underside, the blue-grey back and beak.

It is a bird. An upside-down bird.

CHAPTER 7

The Cabin by the Stream

21 May, 4.35am. I slept last night in a tiny thatched cabin at the top of a garden in Ashbury, Oxfordshire, and have just woken to – or been woken by, I'm not sure which – the local dawn chorus. The cabin is nestled underneath mature trees in a semi-wild area, and as I open the door to better hear the birds, I slowly become aware of the sound of a stream below the garden, running through what used to be a watercress bed. It is earlier than I would like to be awake, but what a way to start the day.

Unlike the birds, the stream has not been to sleep, and it, too, has a song to sing. How can I possibly describe the song of the stream? Simultaneously complex but beautifully simple; 'of the moment' whilst continuously in motion; a bit like a never-ending carnival procession, in that, depending on how, when, and where you hear it, or which way you turn your head, you either feel the full journey of its flow or merely catch a fleeting snippet before it carries on. It seems to me that it is the song of a traveller. Does that make sense? I'm not sure, but know I want to explore this idea further.

I open the door wider. There is a lull in the birdsong now, so I am better able to tune in to the undersong of the stream. There must be some kind of fall, because I can hear the sound of water cascading over rocks, amplified against the silence and the stillness of the night. It is soothing, almost meditative. I sit for a few minutes, tuning the stream's song in and out at will. I wonder, if I were to record the sound now, and then again later, would I be able to tell the difference? Does it sound the same in the middle of the night as it does in the middle of the day? In the middle of winter as in the middle of summer? Rainfall, together with wind speed and direction, will surely make a difference, in the same way that a tune, played by an orchestra with fewer (or more) violins, or under a differ-ent conductor, takes on different hue and timbre. The stream's song is probably softened at this time of year by the leaves in the trees, but in midwinter, when the trees are bare, and the sounds less filtered, I imagine it would sound much different.

I would give the world to live in a place like this, where I might go to sleep at night and wake up in the morning to the sound and dynamism of living water. I love the dynamic power of waves as they crash onto rocks in a storm, and the almost tangible sense of life force as waterfalls cascade from mighty heights into plunge pools below. I am equally happy sitting with my back against a rock, eyes closed, listening to the trickling of a spring seeping from its source on the side of a hill. So often have I done this, that I can now conjure up the sounds of the springs of the Malvern Hills, without even closing my eyes. And when I do close my eyes, the sound becomes even more vivid. It's curious how we can close our eyes to better 'hear' something, and even more curious that the memory of a sound can be enhanced this way.

Whilst pondering the wonder of nature's soundscapes, I remember a short documentary about the wind that I watched not long ago. The film followed birdwatcher and BBC radio producer Tim Dee 'on a walk along the vast open marshland of the southern shore of the Wash, as he embarked on an idiosyncratic mission to capture the elusive sound of "pure" wind.' The programme was a breath of fresh air. Unlike most things on mainstream TV, it was stripped bare: no high-energy presenter, no banter, no clutter, no noise; just a man, walking the flatlands of the Wash and chasing the sound of the wind. There were even great chunks where the commentary ceased altogether so you, too, could contemplate the silence. It was so blissfully daring. It made me hold my breath, hanging on Tim's every occasional word, and also on the momentous spaces between those words. If there is an aural equivalent to the negative spaces in art, then this was it.

I return from my reverie to the soundscape around me. Listening to the birdsong, I instantly recognise it; it is the same as yesterday morning. Same birds, singing in the same trees, at the same time of day. I listen carefully, filtering out first one, then another, until I can tune in on each individual song, like following one instrument in a symphony, or one voice of a choir singing a four-part harmony. But apart from robin and blackbird, I cannot put names to any of the voices singing in this avian choir. I sing along with one or two of the birds, trying to commit the sequences and cadences to memory, in the hope that I might manage to find and identify them online when I get back home this evening.

I know what they are not, which is at least a start. I can confidently say they are neither chaffinch or willow warbler (whose songs I sometimes mix up), nor are they goldfinch, greenfinch, song thrush, sparrow, starling, or cuckoo. Do any of these birds even take part in a dawn chorus? I'm embarrassed to admit I do not know. My son bought me *Collins Bird Songs & Calls*, with accompanying CDs, for my birthday a few years back, but I haven't found time to listen to them properly yet. My birdsong-recognition skills remain extremely basic.

Still, I delight in those birdsongs I do know. I have a similar feeling each time I recognise a bird by its song to that I experience when I overhear someone speaking in a foreign language and realise I understand what they are saying: It is the beginning of a connection. Of course, the birds neither know nor care that I have recognised them, but I know it, and somehow that moment of knowing establishes a great sense of belonging. It is this sense of belonging that I long for beyond all other longings, for it brings with it a feeling of peace so deep and profound that time stands still. It leaves me with no need of, or interest in, the trappings and distractions of everyday life.

These feelings of belonging were probably commonplace amongst our early ancestors, and it is a loss to us that they are today so rare. But given all the noise of the modern world, it is not surprising that we often have a hard time hearing nature. Can you imagine what it would be like for someone whose only auditory experiences had been the natural noises made by other living creatures (including their own kind), together with the sounds of water, wind, and rain, to find themselves catapulted into twenty-first-century London or New York, with their array of alien and discordant sounds?

It is almost impossible to imagine. I suspect the nearest most of us might get to such a scenario would be in learning about the lives of people who have been displaced from their homelands. I recall, for instance, the journey of four siblings from a Sudanese refugee camp in Kenya to Kansas City that was captured in the film *The Good Lie*. You can feel how disorientated they are, how difficult they find it to adapt to this alien environment, so many worlds apart from the place in which they had grown up.

Our great-grandparents would probably had found themselves similarly bewildered had they been catapulted from the nineteenth century into the twenty-first. When we look back ourselves at how previous generations lived, even in the not-too-distant past, we cannot imagine how they survived without the modern conveniences and luxuries we enjoy today. Some of us, myself included, would give the world to go back in time to those days; others would not swap what we have for all the tea in China. I am not suggesting that we try to live without any of the technological advancements that make our lives so comfortable, but I cannot help wondering if our lives are now so full of advancements that we have left no room or time for nature. Perhaps we all suffer, to some degree or another, from a subtle form of displacement.

I know next to nothing about DNA, but I imagine the human psyche might be 'hardwired' to live and thrive in a natural environment not unlike that from which our distant ancestors first emerged. If this is the case, our DNA may not have had enough time to adapt to the rapid changes we humans have collectively brought about over the past two to three centuries, and which we are now relentlessly exposed to. As a result, we are slightly out of kilter.

Maybe the dark, empty holes so many of us talk about when we are feeling low or depressed, the chasms we try to fill up with chocolate, junk foods, alcohol, drugs, and fast living; maybe these emptinesses are nothing more than the absence in our lives of the sights, smells, and sounds that abound in the natural world. I don't know. But what I *do* know is that the moment I step outside and sense the sun on my cheeks, the rain on my hands, or the wind in my hair, I feel more *alive*. These are clichéd examples, I realise, but that's how fundamental and essential the feeling is.

Experiencing the elements makes me feel whole. So, too, does lying on my back amongst the tall grasses of a meadow; strolling through a woodland and pausing to touch the bark of an ancient tree; or taking off my shoes, sinking my toes into the sand, and wading into the sea. Sun, rain, wind, and earth make me feel alive.

Those of us who are fortunate enough to have lives or jobs that involve working outside – like my husband, Rob – are often more in tune with the natural sounds and rhythms of the Earth than those of us who live and work inside air-conditioned or centrally heated buildings. Rob has been a full-time gardener for most of his adult life, so it doesn't surprise me that he is able to recognise so many different birds by the pattern of their flight alone. He has observed, year after year, the birds' comings and goings, and thus knows when each migratory species arrives, where they build their nests, and when their young fledge. He also knows what times of day the resident flocks of finches and tits tend to visit the feeding stations he keeps so well stocked. He has absorbed all this knowledge by some kind of osmosis, in much the way young children learn from what is going on around them without purposefully studying.

I would love to know the birds and their behaviour as naturally as Rob does, but it takes time, and like forging new friendships, it cannot be rushed. I am in awe of those who have made the observation of birds their life's work and passion, and can therefore tell at a glance, simply by its general appearance, that one individual bird amongst a huge flock of seabirds and waders is a rare and exciting migrant from the Americas or the Baltic. I still struggle to tell one gull from another, but I am getting there.

For now, I take pleasure in recognising the sounds of certain bees. It is not so easy to tell a bee species by the buzzing sound it makes, as it is a bird by its song or call, but some are more distinctive than others. Flower bee species, for instance, have a high-pitched, slightly frantic hum; and Shrill Carder bumblebees (so named because of the shrill sound they make as they fly) can be identified as much by their buzz as they can by their appearance, if not more so. I love sitting on the allotment, closing my eyes, and trying to guess from the humming sounds around me whether our visitors this day are hoverflies or bees, and, if bees, whether they are queens or workers. Queens are much larger than workers and their buzz is much deeper.

Sometimes I am startled from my musings as the humming sound around me suddenly switches to a much higher frequency. This tells me that one of my bumblebee visitors has found our tomatoes.

Tomatoes, along with a surprising number of other flowering plants, do not produce nectar, but only pollen, which in most cases is hidden inside an extremely inaccessible, tube-like anther, called a *poricidal* anther. The tightly packed pollen contained within poricidal anthers can be released only by those bees that have mastered the art of buzz pollination, or *sonication*. Honeybees and most solitary bee species (other than carpenter bees and a few others) haven't a clue how to get to the pollen in these plants. Bumblebees, on the other hand, know exactly what to do. As soon as a bumblebee lands on the flower, she grasps the poricidal anthers with her legs, contracts her flight muscles as she would do when warming up or brooding eggs, and vibrates them rapidly, up to twice the frequency of a normal wingbeat. This causes the pollen to explode out of the tips of the anthers and covers the bee with a dusting of nutritious, rich protein.

The sound bees make when sonicating is quite distinct, reminding me of the sound of a high-pitched dentist drill. Not so relaxing as their gentle humming perhaps, but wondrous when you know what it signifies.

Apparently, it is not only humans and other creatures with ears that can hear the sounds of bees and other buzzing insects. Recent research, based on the observation of evening primroses, shows that these plants themselves respond to the sound of bees buzzing. Within just minutes of sensing the sound of a nearby bee, the concentration of sugar in the nectar produced by the plants increases by an average of 20 percent. Incredibly, the flowers even seem to be able to filter out irrelevant noises, such as the wind. This more than validates my playing Mozart to the plants in our greenhouse.

There is something deeply reassuring about the sound of bees buzzing as they go about their business in a flower bed. Most people, even those who are afraid of being stung, would include the 'buzz of bees' amongst the sounds they most associate with summer.

The sounds made by birds, bees, and other living creatures in a habitat are collectively known as the *biophony*. But biophony is just one of three components of the sounds perceived by humans. The other two are *geophony*, which includes non-biological sounds such as those made by waterfalls, wind, or rainfall, and *anthrophony*, the conglomeration of noises created by humans – everything from the buzz of chainsaws in a woodland through to instrumental music and road traffic. Together, these three components make up our *soundscape*, the complex interplay of aural sensations we are exposed to, whether we are sitting in our home or place of work, or walking, miles away from the nearest human habitation.

The sounds I heard when I woke up in the cabin by the stream in Ashbury were all 'wild' sounds, the sounds of the biophony and geophony. The birdsong woke me, bursting suddenly into my consciousness in such a way that it was, at first, the only sound I could hear. As I became more awake and mindful, I was able to discern the songs of individual birds. I noticed the scuttling of some small creature in the undergrowth behind the cabin. I paid attention then to the ambient sounds in the surrounding landscape, such as the stream falling over the rocks and the wind tickling the trees' leaves.

These ambient sounds are the *undersong* of a place. Had I chosen a room in the centre of the village, I might have been woken by traffic, or the sounds of people opening and shutting car doors in a car park. In fact, there was a road not far from my little cabin, but somehow I have become adept at filtering out such anthrophonic noises. When I lived on the Malvern Hills, I was even able to ignore the sounds of chainsaws. I didn't like them, because they made me think of the trees being taken down, so I blocked them out and focused on the song of skylarks instead.

Soundscapes are not only beautiful and elemental; they are also important indicators of dramatic changes being wrought in our ecosystems that we ignore at our peril. In her groundbreaking book *Silent Spring*, Rachel Carson alerted us to the dangers of pesticides and their potential threats to birds and other wildlife: 'Over increasingly large areas of the United States, spring now comes unheralded by the return of the birds, and the early mornings are strangely silent where once they were filled with the beauty of bird song.'

Carson was warning of the dangers of the then revolutionary new pesticide DDT, which was responsible, amongst other things, for crashing bird populations in the 1950s and '60s. DDT was banned in the United States in 1972, but I fear history is repeating itself and we are fast approaching a time when it might be said, 'The summer afternoons are strangely silent where once they were filled with the gentle buzzing of bees.'

The culprit this time is another 'revolutionary' group of pesticides: neonicotinoids ('neonics' for short). They were first introduced in the 1990s and are currently the world's most widely used insecticide. Over 140 different crops – including soy, corn, wheat, cotton, legumes, potatoes, sugar beet, sunflowers, rapeseed, and flax – are treated with neonics.

Neonics are neurotoxins designed to attack insects' central nervous system, causing paralysis and eventually death. They were designed specifically to target agricultural pests such as vine weevils, aphids, whiteflies, Colorado potato beetles, and termites, but the damage they have done over the last two decades to other insects, including bees, has been devastating.

In the past, we were able to 'see' pesticides, since they were sprayed onto crops, and we could wash off many of these contact chemicals. With neonics, though some are sprayed onto crop foliage, the majority are invisible to us, applied as seed and bulb dressings or soil drenches. They are systemic, which means the chemicals are taken up by the entire plant – roots, leaves, fruits, flowers, pollen, and nectar – as it develops.

Previous forms of pesticide tended to be applied reactively, that is, after a pest had been identified in an area, and many fields were never sprayed. For instance, according to the Center for Food Safety, only about 30 percent of corn in the United States was treated. In contrast, neonics are used prophylactically, which means many crops – including between 80 and 100 percent of US corn – are now grown from treated seeds or in treated soil, as a safeguard against the possibility of an attack by the pesticide's target insect. This is a little like taking antibiotics throughout the year just in case you are exposed to somebody with a chest infection come December.

Ironically, neonics were designed to be 'safer' for bees and for the environment. It was believed that they would be more effective than spraying crops with chemicals targeting specific pests, and that they would be more environmentally friendly because the crops wouldn't need repeated spraying. Unfortunately, their sublethal effects were not tested before they were used in our fields.

Because neonics are water soluble, they can remain in the soil for a number of years. Only 5 percent of what is applied to the seeds is actually taken up by the plants. The remaining 95 percent of the pesticides leach into the soil and, because they are water soluble, migrate via the groundwater into nearby streams and waterways. Neonics also get sucked up from contaminated soil – not only by subsequent years of crops grown in the treated field but also by the wild flower margins along the edges of the fields, many of them, ironically, planted or allowed to grow specifically to support pollinators. Margins and waterways weren't intended to be treated with neonics, but they have been, resulting in sustained exposure for a multitude of organisms, including bees, moths, butterflies, hoverflies, and aquatic invertebrates. And there are more, far-reaching effects, too, since declines in these insects cause shortages in the food chain for various insect-eating bats, amphibians, and birds.

When the effects of neonics first began to be noticed, people's concerns were focused on honeybees. This was around the time of colony collapse disorder, and neonics were suspected to be one of the root causes. But for many years now it has been clear that bumblebees and other invertebrates are also being affected. These insecticides are *not* 'safe for bees', despite the reassurances we were given in the 2000s, and for a simple reason: When bees collect pollen and nectar for their brood to feed on, neither they, nor their larvae, can escape the neurotoxins' effects.

Scientific evidence collected over the last decade or so proves that neonics produce numerous chronic symptoms in bees, such as interfering with their navigation systems; disrupting their foraging behaviour and ability to communicate; impairing their immune systems; and reducing their reproductive success. And recent evidence has found that higher levels of neonicotinoid residues in the nests of solitary bees may be linked to their producing fewer eggs cells.

Bees and other creatures have also, in some cases, been acutely poisoned due to mechanical malfunctions during the sowing of neonic-treated seeds. In 2008, millions of honeybees died in Germany when a neonic-laden dust, which had not stuck properly to treated maize seeds, was released from drilling machines. A similar incident occurred a few years later in the US Midwest.

As if these incidents are not worrying enough, researchers have discovered that neonics are poisoning birds, too. Just five neonic-coated maize seeds left scattered on the ground after sowing can prove lethal if ingested by a bird the size of a grey partridge. We are still being told by the manufacturers of these pesticides that they are safe for bees and the environment, much like we were told in the 1960s that DDT was safe for humans. Who knows what else is being affected by these neurotoxin pesticides?

In response to concerns about the lethal and sublethal effects of neonics on wildlife, in 2018 the EU banned three of the most widely used neonics – clothianidin, imidacloprid, and thiamethoxam – from any use outdoors, including on crops and lawns. However, other neonics and newer generations of systemic insecticides are still permitted, and the soil in potted plants is still allowed to be treated with neonics, including those that are otherwise banned.

Also, despite scientific evidence and advice, EU member states have yet to introduce tests to ensure that before a pesticide was approved, its sublethal and long-term toxicity to honeybees and wild bees would be assessed. Until the regulatory system for pesticide-risk assessment changes, we are doomed to repeat history, having, sadly, not learned from it.

Life on Planet Earth is underpinned by the insects, invertebrates, and other small creatures that are now being destroyed in their billions, not just by neonics, but by toxic cocktails of these and other insecticides, fungicides, and herbicides which, when insects are exposed to them, cause more damage together than the sum of their parts. The continued and indiscriminate poisoning of animals and plants, not to mention the water and soil upon which we rely to sustain life itself, is extremely dangerous. So, to my mind, addressing the worldwide use of neonicotinoids and other pesticides is as pressing as any other current environmental issue.

There are some who believe key crops will fail if seeds, plants, and soils are not treated with neonics and other pesticides. It is true that I am neither a scientist nor a farmer, just a person who, like many others, cares enormously about the plants and animals we share this planet with. But whilst it's all very good making a big noise about the possibility of oilseed rape crops failing if neonics are banned, surely the alternative – there being no pollinators left to pollinate the crops – is not viable. Indeed, it doesn't bear thinking about. Instead, we must search for more sustainable ways of feeding the world, starting with growing our crops without the prophylactic use of pesticides. We could begin with an increase in small-scale organic farming. Whatever the solution, the problem needs to be owned and addressed by us all scientists, politicians, growers, and consumers alike – and urgently.

When I was a child in the 1960s, we used to travel up the A1 to Yorkshire to see my grandmother. I remember my father having to make regular stops to wash the windscreen, it was so splattered with dead insects, the wipers alone couldn't keep it clean. I also remember watching huge flocks of birds following farmers' ploughs in the fields alongside the road. Looking back now, I realise they were feeding on the abundance of worms and other organisms, living just beneath the surface of the soil, that had been exposed by the plough.

These days there are so few insects that our windscreens remain clear from Land's End to John O'Groats, and I only rarely see flocks of birds following the tractors, because in many of the fields there is little or no life left in the soil.

Insects are resilient. They have collectively endured five major mass extinctions on Earth. But the current catastrophic declines in populations, and extinctions of species and families, are different. They are not being caused by ice ages, or meteorites. They are being caused by us.

Nearly sixty years ago, we learned the painful consequences of using DDT from Rachel Carson and her world-changing book. Never has it been more important to listen to – and really hear – what the natural world's soundscapes are telling us.

CHAPTER 8

Cuckoo, Cuckoo

We all know what a cuckoo is, don't we? It is a bird that lays its eggs in the nests of other birds, relying on those other species to raise its young. A cuckoo lays between twelve and twenty-two eggs in a season, each one in a different nest, usually in a nest belonging to the same species of bird that reared her. In all my life, I have only once seen an adult cuckoo. She was sitting on a telegraph pole in Northern Ireland, and I was quite sure she was a sparrowhawk, until she flew directly over our heads and I was able to see her plumage, which was more striped than speckled. But I have seen many images of cuckoo fledglings being fed by their poor, diminutive adoptive parents. So oversized are the cuckoo young compared with the birds' *own* young, that the cuckoo fledglings rarely fit inside the nest. Indeed they are often larger than the nest itself.

We happily forgive the cuckoo her cheeky behaviour, making allowances because she is our harbinger of spring, bringing with her, when she arrives from Africa, promises of warmer days and sunshine. What matter that she sneaks into the nests of sweet but unsuspecting dunnocks, meadow pipits, and reed warblers, replacing one of their eggs with one of her own, or that her young, when they hatch out, usurp their adoptive siblings, greedily demanding all that the exhausted parents can bring back to the nest, and more. No, all that matters for us is that when we hear the first (male) cuckoo call of the year, we know all is well with the world.

Cuckoo bees, though, are not so easy to love.

The term 'cuckoo bee' applies to a number of different bees that, like their namesake bird, lay their eggs in the nests of other bees. All are brood parasites, or *cleptoparasites*. And just as bumblebee and solitary bee life cycles differ, so do the life cycles of their respective cuckoos.

Cuckoo bees account for around 80 of the 270 or so solitary bee species in Britain and Ireland. Typically, each one is associated with a particular host bee species, but in some cases, they are associated with more than one host bee species. Not all bees need worry about cuckoos. In Britain there are no known cuckoos for Yellow Loosestrife bees (*Macropis europaea*), for instance. Conversely, very few of our ground-nesting *Andrena* species get away *without* being cuckoo'd. The cuckoos of ground-nesting mining bees are mostly of the sort known as *nomads*.

Such a nomad can usually be found scouring nesting sites, in search of a host bee. Once she has located a nest, the female nomad bee hangs around outside, waiting until the hard-working owner of the nest has popped out to collect pollen, then nips in and lays her own eggs in the nest. She will lay eggs in multiple hosts' nests over the course of her reproductive life. When the nomad's young hatch out, they either use their long mandibles to destroy the host bee's larvae, or eat so much of the pollen that the host bee's larvae starve to death. Either way, directly or indirectly, the host bee's larvae don't stand a chance once the nest has been infiltrated by a cuckoo bee.

I have often wondered why certain nomad species are associated specifically with certain hosts. The answer, it seems, lies in the fact that the cuckoo bee species have become so highly specialised that their larvae require exactly the same amount and type of pollen as their host larvae require in order to develop. In fact, nomad bees and other cuckoo bees are so co-evolved with their hosts that they no longer have any pollen-collecting apparatus of their own. At this point in time, it would be impossible for them to exist without their hosts. The cuckoo bee species, in many cases, are actually far rarer than the hosts.

When I am out and about searching for ground-nesting solitary bees (*Andrena* species), I am often first alerted to the location of a nest, or an aggregation of nests, by the presence of nomad cuckoo bees. This is because the cuckoo bees that lay their eggs in the nests of ground-nesting bees are easy to spot. They are often brightly coloured, mostly yellow and black, but often dotted with patches of red, and inconspicuously haired, resembling tiny wasps more than bees. As cuckoo bees spend so much time lurking in the open, waiting for an opportunity to zip inside a host's nest, this bright colouration may help to protect them from predators, which more often than not mistake them for wasps. It is not just their colour that catches my attention, but also their behaviour, as they fly back and forth, close to the ground, searching for suitable host nests.

I have to put up my hand and say I find it almost impossible to tell some of these nomad cuckoo bees apart.

At the bottom of Stoney Path, which leads steeply down from Shaftesbury Abbey into Tanyard Lane, and then onto St James, sits a block of garages. These garages are edged by an ancient Shaftesbury greenstone wall, known locally as the 'Bee Wall'. Every spring this wall comes alive as its residents, a large aggregation of beautiful, Hairy-footed Flower bees, emerge, one by one, from their winter diapause. The ginger-coloured Hairy-footed Flower males appear first, usually sometime in mid to late March (though, with the warming associated with climate change, they sometimes come out earlier), followed a week or two later by the all-black females. Matings take place, pollen is collected, and the females settle into the business of creating new nests inside the wall, carefully rearranging the mortar to create individual cells, before providing each cell with exactly the right amount of pollen, laying an egg, and sealing it.

Along with other Shaftesbury residents (of the human kind), I love to stand by the Bee Wall watching all the comings and goings, appreciating the beauty of these pretty little bees, and wondering how the nest-building females manage to put up with the males' attentions, when all they want to do is get on with laying their eggs and collecting pollen.

It has been raining this morning, and as I walk through Tanyard Lane on my way up to the Thursday market, I notice a moustachioed male sheltering inside one of the crevices in the wall. Being careful not to frighten him, I gently crouch down to get a closer look. He is extremely handsome, with an unusual, blunt-looking yellow masked face. Hairy-footed Flower bees rarely stay still for long, so it is a treat to watch him. Then, at once, the sun is out, and the Bee Wall is humming with activity. When I turn back to the crevice, he has gone. Now, instead, females are bringing back pollen from nearby comfrey, wallflowers, and lungwort.

It is whilst I am watching a particular female, who doesn't seem to be able to find her nest entrance, that I spot another bee, lurking in a suspicious manner just above the entrance to some other flower bee's nest. This is a sinister-looking character if ever I saw one: a dark grey that is almost black, with silvery white dots along the sides of its abdomen and creamy white patches of hair on its legs. Do Hairy-footed Flower bees have cuckoos? I do not know, but if they *do* I have a strong feeling this bee might be that cuckoo. I take a photograph and whizz up to the market, before heading home to see if I can discover its identity.

Back home, I consult Steven Falk's *Field Guide to the Bees of Great Britain and Ireland* and confirm that my unknown bee is, as I thought, a clepto-parasite of the Hairy-footed Flower bee – the Common Mourning bee (*Melecta albifrons*). However, I cannot tell for sure if the bee I saw was male or a female, as both look quite similar. As it was lurking outside the nest entrance of its usual host, I guess it was female.

Apparently, this cuckoo bee is 'quite sporadic and unpredictable in occurrence', so I am delighted to have discovered it living in Shaftesbury. Or am I? After all, I know exactly what she and her kind are up to, and cannot find it in myself to approve of her behaviour. She is a brood parasite, existing and reproducing at the expense of her host species – in this case, my beloved Hairy-footed Flower bees – whose larvae she destroys to give life to her own. She is a murderess, no less. Not for the first time in my journey back to nature do I have to remind myself to take a step back and not judge wild creatures as we judge other human beings. There are far more Hairy-footed Flower bees than there are Common Mourning bees around. With this thought in mind, I decide to adopt a laissez-faire attitude to the cleptoparasite.

Well into May, the Bee Wall remains a hive of activity. Now that the females are all mated and busy filling their nests, the male Hairy-footed Flower bees are nowhere to be seen. They have all died by now. But the female bees, and their cuckoos, continue flying, in and out, into early June.

I notice the Common Mourning bees are extremely partial to the patches of green alkanet growing in the garden opposite the wall. Maybe this is one of the reasons we have a healthy population of them in this part of Shaftesbury. Who knows.

By mid June, the Bee Wall is quiet. Deep inside, there are hundreds of developing bees. What proportion, I wonder, are cuckoos?

In the same way that cuckoo solitary bees are each associated with a specific host bee (or bees), so, too, are cuckoo bumblebees associated with particular bumblebee species. The Vestal cuckoo takes over the nest of the Buff-tailed bumblebee, and the Forest cuckoo takes over the nest of the Early bumblebee, and so on. There are just six species of cuckoo bumblebee in Britain and Ireland .

As we approach the end of July, I am happy to note that this has been a really good year for Red-tailed bumblebees (*Bombus lapidarius*). I hardly saw any last year, but there now must be a number of locally active nests around town, because I have seen as many of this species as any other bumblebee, perhaps more. The nests have clearly been successful, because some queens and males have already emerged, which makes me extremely happy. I have also seen no fewer than five Red-tailed Cuckoo bees (*B. rupestris*) this year, more than I have seen altogether since I first learned to recognise them.

Female Red-tailed cuckoo bees are magnificent. They are the largest cuckoo bee in Britain and Ireland, and have the greatest average wingspan of all our bumblebees. As their name suggests, they closely resemble their host queens, the Red-tailed bumblebees, and you might expect that differentiating between the two would be a challenge, but it is not. There is something about this bee, something about its demeanor, not to mention its buzz (which sounds like a cross between a Chinook helicopter and an angry queen hornet), that says *I am* Bombus rupestris: *Be afraid; be very afraid.* If I were a Red-tailed bumblebee queen, I would be afraid. Absolutely I would.

It was towards the end of May that I saw the five Red-tailed cuckoo bumblebees, three of them on the same day, foraging on dandelions in St James's Park, the others a week or so later on the other side of town. Cuckoo bumblebees always appear, to me, to be either falling asleep, or just waking up. They never seem in a hurry to go anywhere, so when you see them you usually have time to go back home, boil the kettle, make a flask of tea, and collect your camera or sketch pad. When you come back they are likely to be still there, probably on the same flower. These were no exception. Two of the three I found in St James's Park were still there nearly twenty minutes later, and each posed beautifully for me, looking even more striking against the bright golden yellow of the dandelions than they usually do. Like the bee whose nest she will soon attempt to take over, the Red-tailed cuckoo bumblebee's head, thorax, and abdomen are jet black, and her tail bright red. Her wings, though, are noticeably darker than those of her host species. At first they appear to be blackish brown, but when they catch the sunlight, this bee's wings shimmer with dark, luminous blues and purples.

Apart from their lethargic behaviour and (mostly) darker wings, the most noticeable difference between the six cuckoo bumblebees and true bumblebees, is that the cuckoos lack pollen baskets. Like solitary cuckoo bees, these bees rely completely on other bees to rear their young, so have no need for their own pollen-collecting apparatus.

Cuckoo bumblebees also have less hair than true bumblebees. In fact, there are areas of their bodies where the hair is so sparse that you can see the cuticle, hard, black, and impenetrable, almost as though they have been endowed with body armour, which, in a way they have. Like all other invertebrates, bees' inner organs are protected by an *exoskeleton* made from chitin. The outer case of the shiny, black chitinous armour we see beneath a bee's body hair is known as the *cuticle*, and in the case of cuckoo bumblebees, this hard shell-like exterior is thicker than it is in true bumblebees – all the better for resisting stings.

Two other physiological differences between cuckoo and true bumble-bees are also related to the cuckoos' need to be less vulnerable and more formidable in battle. First, true bumblebees have gaps between their abdominal segments, from which they secrete thin slithers of wax to use in the nest, whilst female cuckoos don't have these breaks in their defences. Second, cuckoo females are equipped with stronger mandibles and longer, more curved stings than true queens and worker bumblebees. The sum total of these adaptations makes female cuckoo bumblebees fearsome opponents. When they go into combat, fatalities are likely to occur on the other side.

Cuckoo bumblebees usually emerge from hibernation later than their host species. After feeding to replenish their energy levels and develop their ovaries, they are ready to lay their eggs, and it is at this stage that they can often be seen patrolling grassy banks and other likely areas in search of established host nests. It is believed that cuckoo bumblebees use their antennae to locate the nests of their unfortunate host bees via scent trails unwittingly left by worker bees as they travel between the nest and patches of flowers.

Once our fiendishly clever cuckoo has located a suitable nest, she will sometimes hang around the entrance for a while before sneaking inside. At this stage, the colony will usually attempt to repel her. If she is not noticed, the cuckoo might lurk quietly in a corner of the nest, gradually acquiring the scent of her host whilst she waits for the moment to make her move and usurp the incumbent queen. When the time is right, the cuckoo bumblebee will attack and kill, or subdue, the true queen. This attack rarely goes undefended, and many workers can lose their lives in defence of their queen. If the cuckoo bee is victorious, she will do a quick sweep through the nest to kill any existing eggs or young larvae, and then lay a big batch of her own eggs. She forces the old queen's workers to submit to her rule by using a combination of aggressive behaviour and pheromones.

Some female cuckoo bumblebees seem to come and go as 'visitors' without attempting to take over the host's nest. I witnessed this myself when I was called to an Early bumblebee nest that had had a very lucky escape from a garden bonfire. A friend had started the fire to burn a pile of garden debris when he heard frantic buzzing arising from beneath some of the leaves at the edge of the pile. He quickly doused the flames, before any great harm was done, but the bumblebees had to spend the next few days repairing what was left of the now-exposed nest. I watched them working, dragging dried leaves and tiny twigs across the ground to cover their nest. It was whilst I was observing these repairs that I noticed a huge female cuckoo land nearby.

She clambered in and out of the nest at will, sometimes walking right over the Early bumblebee queen, without the queen or her workers paying the slightest attention to her presence. Maybe they were too busy fixing their home to stop and pay her heed, but the cuckoo did not seem to pose a threat. Interestingly, this was a Vestal Cuckoo bumblebee, which usually takes over the nest of a Buff-tailed bumblebee, so what she was doing in an Early bumblebee nest, I do not know. It appears, however, that this behaviour is somewhat common, and that she might just have been using the nest as something like a free B & B.

Cuckoo bumblebees do not produce their own workers, so rely entirely on the colony's existing workers to raise their young. The timing of an attack, therefore, is absolutely critical, with one of the most important considerations being the stage the host colony has reached. On the one hand, if there are too few workers in the nest, the cuckoo is less likely to be challenged, but if she kills the queen at this stage, there may not be enough workers to raise her own brood. On the other hand, if she waits for more workers to be produced and there are too many workers when she attacks, she risks being overcome when they come to the defence of their queen. The cuckoo bumblebee's success is by no means a given.

Suffice to say, if the cuckoo does manage to usurp the true queen, the nest is then extremely unlikely to produce its own reproductives. There are exceptions to this rule, where both the true queen and the cuckoo somehow manage to live alongside each other, with the workers of the true queen rearing both broods, but my understanding is that this is unusual. In most cases, the cuckoo, as new ruler of the roost, is capable of laying many more eggs in each batch than the poor deceased or defeated queen was able to. When these cuckoo eggs hatch, the larvae are all fed and reared by the colony's workers. Whilst there are males and females, there is no worker caste, and no 'queens' as such, amongst these cuckoo young.

There are usually more male cuckoos than females, and the males can often be found clustered in groups on flower heads. I have seen such groups gathered together on *Knautia*, knapweed, and various thistles – and like the female cuckoo bumblebees I saw in St. James's Park, they were extremely lethargic, and therefore, extremely easy to observe.

Once mated, the fertilised females go into hibernation. The males fade and die before winter sets in. The following spring the female cuckoo bumblebees emerge, fresh from their winter sleep and ready to commence the whole life cycle again.

Until a few years ago, I could have summed all I knew about bumblebees in just a few short sentences, and here I am talking about 'host queens', 'cleptoparasites', and 'slithers of wax secreted from between abdominal segments'. What started out as a concern about bee decline has developed into an all-consuming interest in and passion for these creatures, to the extent that I am prepared to trawl through scientific papers which speak to me, for the most part, in double Dutch, so I might further my understanding of this aspect of bee evolution.

When I first learned how cuckoo bees behave towards their unsuspecting hosts, I have to admit I didn't think very kindly towards them. And it appears I was not alone. In F.W.L. Sladen's *The Humble-Bee*, published in 1912, the summary given of a cuckoo bumblebee is heavily weighted in one direction:

> *The way in which the...[queen] proceeds to ensure the success of her atrocious work has all the appearance of a cunning plan, cleverly conceived and carried out by one who is not only mistress of the crime of murder, but also knows how to commit it at the most advantageous time for herself and her future children, compelling the poor orphans she creates to become her willing slaves.*

But if you were to ask whether cuckoo bees pose a threat to the existence of their host bees, I would say, on balance, they do not. It is surely not a coincidence that the year I saw more Red-tailed cuckoo bumblebees than ever before also happened to be a year I saw great numbers of their host bees, the Red-tailed bumblebees. Without a healthy host population, you cannot have a cuckoo population, healthy or not. It is not in the interests of the cuckoo bees to wipe out their hosts, for without them they cannot themselves exist. If a particular bee species happened to be on the very brink, I imagine a sudden influx of cuckoo bees might conceivably push that species over the edge. Otherwise, nature, when left to itself, has a way of maintaining balance.

CHAPTER 9

On Swarms and Stings

Between late spring and midsummer, my inbox, mobile phone, and social media timelines fill up with messages from people asking for advice about bee swarms. As Rob is a beekeeper, my first thought when I see the word 'swarm' is the process by which a resident queen honeybee leaves her hive, accompanied by thousands of her loyal worker bees, to search for somewhere suitable to set up a new colony.

So, if it is clear that the enquirer has seen a honeybee swarm, I typically suggest contacting the local beekeeping association, which will more often than not send a swarm collector out to them. However, if the enquirer is local to us, Rob himself will often jump in the car with his bee suit and an old wicker laundry basket to see if he can catch the bees and bring them home. Beekeepers love nothing more than to retrieve and rehome a swarm of honeybees, and if we don't have a spare hive ourselves, there are always other beekeepers in the area who are delighted to take the swarm off Rob's hands.

Taking a container of some sort to catch the bees is a necessity, as is making sure your container has a lid that closes properly for the journey home.

Wearing a bee suit, though strongly advisable for anyone inexperienced, is not. Some beekeepers tend to their bees without protecting themselves in any way whatsoever; the wonderful Heidi Herrmann, president of the Natural Beekeeping Trust, has a video online in which she catches a mid-summer swarm, gathering them up by hand into a skep (a wicker beehive), before gently coaxing them into one of her empty hives – all whilst wearing only her light summer clothes. This takes a leap of faith and a lot of trust on the part of the swarm catcher, and is certainly not for everyone. In Heidi's case, the swarm was from one of the hives she keeps in her garden, so she already knew the bees and they knew her.

Heidi had not always been this comfortable around bees. Indeed, a fear of being stung was one of the first obstacles she needed to overcome when she took up beekeeping around twenty years ago. She approached this slowly, starting to leave off items of protective clothing such as gloves and veil, until eventually she was able to open a hive without any protection at all. Heidi describes the moment as 'a kind of unity'.

Rob prefers to wear a suit, especially when dealing with bees of unknown origin.

Honeybees are actually less likely to sting whilst swarming than at any other time. This is because, prior to swarming, the bees gorge themselves on honey, so are more docile than ever. They are also so full up, and their honey stomachs so distended, that they are less easily able to curl their abdomen under to sting.

Some swarms are easier to collect than others, involving nothing more than gently brushing or shaking the swarm into a container. Ladders are occasionally needed, as are screwdrivers or other tools, if the swarm has decided to try to make a new, permanent home inside (for instance) a dovecote or underneath some floorboards.

The most difficult collection Rob has ever undertaken was not a swarm, but a wild colony that had some years ago established itself between the outer and inner walls of Sutton Waldron village hall. The village hall committee had asked other beekeepers in the area if they could help, but there were no takers, possibly because a cursory inspection revealed that the outer, wooden wall of the hall would need to be completely dismantled to reach the colony. This would be extremely time-consuming, and catching the queen – essential to the survival of a colony – would by no means be guaranteed. Workers were coming in and out of the hall in ever increasing numbers, and as the space was used regularly for community activities (including a mother-and-toddler group), the committee had reluctantly determined that the bees would have to be destroyed if they couldn't be removed. Rob stepped up as their last hope.

Rob loves nothing more than a challenge, so over the next four weeks, he slowly, painstakingly, and extremely gently dismantled the outer wall of the hall, exposing an enormous honeycomb that spanned a number of the joists and extended up into the roof. The colony had clearly been living here for quite some time, and had grown so huge that it could no longer be contained inside the wall – which was why people were getting more and more bothered by the bees.

Rob quickly worked out that the colony was so large, it would need to be removed in two parts. He had attached a tarpaulin to the side of the building whilst he worked, because although the weather was warm, it was also very wet, and the exposed comb needed to be protected between his visits. Then late one Saturday evening, after warnings of upcoming storms and heavy rains, Rob had to commit to removing those bees that he could access, and just hope the queen would be amongst their number. Aided by one of the villagers, he took a knife and cut off the exposed part of the comb, complete with brood cells and many thousands of worker bees, allowing it all to drop into a cardboard box.

Only when he started to cut the comb away could he see that the bees had built their comb around the nails that held the outer casing of the hall to the inner casing. As the bits that were not attached to the nails fell into the box, he realised the queen was probably amongst the tightly clustered group of bees that had made their way up into the roof whilst he had been working.

Rob doesn't give up easily. Against the clock, he made a long, thin makeshift box, tucked it around the comb remaining just beneath the roof space, and left it there overnight. He hoped the queen and those workers he hadn't already caught might crawl back down to this comb, which was still attached to the side of the hall but also now contained in his makeshift box.

We came back again late the following evening so that Rob could prise the comb inside the makeshift box away from the wall. This final part of the rescue – for a rescue is what it had become – was executed with great precision in pelting rain. Sometime shortly before midnight, we drove home, exhausted but jubilant, with tens of thousands of bees and giant chunks of honeycomb divided between two boxes in the back of the car.

When we got home, Rob tipped the contents of both boxes into an empty top bar hive and put the lid on. We would not know until we returned to the hall the next day to retrieve stragglers whether the rescue had been successful or not. Happily, when he did go back up the ladder and into the roof space, Rob could see no sign of anything that even slightly resembled a cluster; there were less than a handful of confused, disorientated bees up there. What relief. It looked as though he had managed to collect the queen and pretty much the entire colony of workers, together with most of their comb, in the nick of time. The worst of the stormy weather that had been forecast arrived that afternoon.

———

Three years on, we call this colony, which survived against the odds, our Village Hall Bees. They have adapted well to their new home, thrown off at least two swarms of their own, and hold a special place in our hearts.

But not all swarms during the swarming season are honeybee swarms. If ever there is mention of bees swarming around the entrance of a bird box, or under the eaves of someone's roof, and if the swarm appears to be made up of just *tens* of individual bees rather than *thousands*, then I can hazard a pretty good guess that it is made up of Tree bumblebees.

Tree bumblebees (*B. hypnorum*) are the new bumblebees on the block. It is not entirely known how this species, which is already widespread in mainland Europe and parts of Asia, first arrived in Britain and Ireland. It is thought they might have been blown across the English Channel from northern Europe, or arrived as hibernating queens in the soil of imported plants. Either way, since they first appeared in the New Forest back in 2001, they have spread rapidly.

So successful and so fast has been their spread that their range now stretches as far north as Scotland and west across the Irish Sea. Their progress is being monitored by BWARS, and thus far, they do not appear to be affecting our native species in any way, so they are not considered to be an invasive species. The blue tits, whose boxes they appear to have a particular penchant for, might dispute this, especially those that have been evicted in the early stages of their own nest-building by a Tree bumblebee queen looking for somewhere to start her colony.

The success of the Tree bumblebee, whilst many other bumblebee species are struggling, is probably down to its unique approach to nesting. Most bumblebees prefer to nest at ground level, often in abandoned rodents' nests, under compost heaps, or in tussocky grass, but Tree bumblebees choose hollow trees – hence their name – and bird boxes, preferring those in which a bird has previously nested. As we British are a nation of bird lovers, and as no self-respecting wildlife gardener would be without at least one bird box in the garden, there is no shortage of suitable nesting habitat for this opportunistic species.

Nor is there much competition from other bees when it comes to these unusual nesting sites. It is no wonder, then, that Tree bumblebees are doing so well. In fact, it's not just bird boxes they like; this species regularly nests under the eaves of houses and in other man-made cavities, such as tumble dryer vents and postboxes. One downside to these locations, which are so easy to find, is that they seem to leave Tree bumblebees particularly vulnerable to attacks by wax moth caterpillars.

Because Tree bumblebee nests are often at or above eye level, they are also easy for us to find and watch. A few years ago, I watched, fascinated, as a Tree bumblebee queen established her nest inside an old cannon in Diana's garden that had previously been home to a family of great tits. That same summer, David, who farms nearby at Berrybrook Farm, in Sedgehill, showed us a thriving colony of the bees inside an old mouse nest nestled on the side of his slurry tanker.

Amazingly, during the colony's three-month residency, David continued to drive his tanker out to the fields to spread slurry at least twice weekly, before parking it back up in the evening, in exactly the same spot just inside the gate of the field across the road from his farmyard. It must have been confusing for the worker bees to return from their foraging trips, laden with pollen and nectar, to find their home temporarily missing!

The thing I love most about the story of this colony is that, although David was stung once or twice when he disturbed the bees to connect the slurry tanker to his tractor, it did not once cross his mind to turf the bees out of their home. If only more people could adopt such a 'live and let live' attitude, as David does.

Aside from the telltale clues revealed in their choice of nesting location, Tree bumblebees happen to be one of the easiest bumblebees to identify visually, having a distinctive colour banding that comprises a ginger thorax, black abdomen, and pure white tail, a bit like a tricolour flag. Their ginger thorax often gives the appearance of having a slight bald patch in the middle of it. Unlike some bumblebee species, the queens, workers, and males of this species do not differ much in appearance, so it is not easy at a glance, to tell them apart. The queens are larger than the others, however.

Having said how easy they are to identify, I will concede that Tree bumblebees do occasionally throw up variants and aberrations. Some are almost completely black, others mostly black with a white tail. I found an all-black male sleeping inside a calendula flower on our allotment this year. That is, I thought it was a male, and encouraged 'him' to climb onto my hand so I could get a closer look. It was only later, when I posted a photograph on Twitter and Steven Falk pointed out that 'he' was a 'she', that I realised my mistake. Had I known this at the time, I would not have been so blasé.

Calendula is not one of the flowers I usually associate with Tree bumblebees, though I have seen other short-tongued species foraging on it. Tree bumblebees show a preference for flowers in the Rosaceae family, such as brambles, raspberries, and cotoneasters. They also seem to like flowers that hang downwards, such as heather, snowberry, and *Cerinthe*. By far the plant they seem to enjoy most on our allotment is the comfrey.

If it becomes apparent that the swarm I am being contacted about is a Tree bumblebee colony, I try my utmost to convince my enquirer to leave it alone. For the colony to have reached the stage where you might say you are seeing a 'swarm', a nest must already have been in place for at least a couple months. The sudden increase in activity indicates that the colony has produced new daughter queens, and by giving off some kind of scent or pheromone, these newly hatched queens are attracting all the males from miles around. These males, usually from other colonies, congregate outside the nest, dancing frantically around the entrance and bumping into one another, waiting for the young queens to emerge and, for all the world, giving the appearance of a swarm.

The important thing to know is that this pre-mating stage will not last long – a matter of days, perhaps, and certainly no more than a week or so. If allowed to get on with their lives without any interference, and preferably without the attentions of a pest controller, the daughter queens will very soon leave the nest. As soon as they do, the waiting males will pounce on them and mating will take place. This is the only common bumblebee species in Britain and Ireland where the males employ this tactic, rather than hanging out in gangs at some 'rendezvous station' for the queens to come and find them.

I wonder if Tree bumblebees are able to adopt their tactic *because* of their habit of nesting in tree hollows and the like. Bumblebees that nest beneath the ground, sometimes in tunnels up to five metres long, must have a much more difficult time, with so much soil separating them, and their pheromones, from the outside world. Or maybe the Tree bumblebee queens just give off stronger pheromones than other bumblebees. Unlike other bumblebee species in Britain and Ireland, Tree bumblebee queens are polyandrous, which means they are likely to mate more than once. They also have the shortest copulation duration of any British bumblebee.

Once done with mating, like our other bumblebees, the new Tree bumblebee queens stock up on nectar and go into hibernation until early the next spring. The old queen, together with all the remaining workers and the dancing, pouncing males, will not survive for longer than a few more weeks. So, when you catch sight of a 'swarm' of Tree bumblebees, it is a time to enjoy and celebrate the fact that you have provided a home for these most interesting creatures to complete their life cycle.

A final word of caution is due, however. The female guard bees are on high alert during this last stage of the Tree bumblebee nesting cycle, and can be a little more defensive (and occasionally aggressive) than usual in their efforts to protect the new daughter queens – not just from potential predators, but also from the unwelcome attentions of the males before they are ready to leave the colony and mate of their own volition. My advice is to steer clear of Tree bumblebee nests whilst the males are dancing outside.

I say this with the painful benefit of hindsight. I once came across an active Tree bumblebee colony nesting in the hollow of an old elder tree on the banks of the River Severn in Worcestershire. This was soon after their first appearance in England, and it was quite an exciting find. I began to narrate a short video of the dancing males to capture the moment. 'This is a Tree bumblebee nest,' I began. 'As you can see, these bees are not aggressive. They are completely undeterred by my presence —' at which point, an incoming worker bumped into my head and became tangled in my hair.

I am ashamed to say I reacted badly, shaking my head and making quite a lot of fuss. This must have alerted the guard bees, who flew straight at me, buzzing me furiously as I ran as fast as I could away from the nest. Still, they chased after me, and I was stung twice in my left ear; I ended up with a cauliflower ear any rugby player would have been proud of. But looking back on the day's event, I feel I deserved to be stung by those bumblebees. I should have shown more respect to their nest and life cycle, and I should have been more aware that I was standing in the flight path of the incoming and outgoing foragers. The guard bees were only doing their job, defending their sister, who presumably sent out distress signals as soon as she became entangled in my hair. Had I not reacted so foolishly, I might never have been stung. Needless to say, I didn't upload this particular video to YouTube.

Only in rare cases will a bee sting you without being provoked. Indeed, most species of bee don't (or can't) sting at all. Bumblebees are normally very peaceful, gentle creatures. Yes, the females are equipped with a sting, but they use it only if they have to; they have far more important things to do than go around stinging human beings. In normal circumstances, you would need to sit on a bumblebee, put your foot inside a boot where one was taking refuge, or disturb its nest, for it to resort to stinging you – and even then it would only be in defence.

It is commonly believed that bees die after stinging. This is true of female honeybees, but not so for bumblebees. Fortunately for bumblebees, their stingers resemble needles, so are easily withdrawn after piercing the skin. If a bumblebee feels threatened by your presence, it will politely signal to say you are too close for comfort and that it would, please, like you to go away. Bumblebees do this by raising one of their middle legs in the air, as though waving at you.

If you back away, the bee will relax and put her leg back down again. If, however, you move closer, and if the bee is unhappy about this, she will lift another leg in the air. If, despite this warning, you go closer still, she will raise both legs up, vertically in the air, and turn onto her back to show you her sting. This behaviour, known as *posturing*, very rarely leads to the bee actually stinging you – it means that she is low on energy, unable to fly away, and would like you to move out of her space, as soon as possible.

Male bumblebees posture in the same way as the females, but as they do not possess a sting, they are just bluffing.

Honeybees workers are equipped with a sting, which they will use if they believe their honey stores, or their queen, are being threatened. They will also attempt to use their sting if they think you are threatening their lives – say, by standing or sitting on them. Unlike a bumblebee, a honeybee worker's sting is barbed, and will remain under your skin after you have been pierced. As she attempts to fly away from the site of the sting, her sting and intestines are pulled out, so unless you can remove her quickly, and without damaging the sting, the unfortunate bee will die. Honeybee queens can sting repeatedly, but as they spend almost their entire lives inside the hive, the odds that you will ever encounter one are fairly remote. Male honeybees, like all other male bees, have no sting.

It is worth noting that honeybee colonies have somewhat variable temperaments, ranging from extremely docile to quite tetchy. This is a result of genetics, with certain crosses being hard to handle, even by experienced beekeepers. The good news is that honeybees almost never sting anyone who is not close to their hive, so you need not worry about being stung whilst out gardening or walking through a field. But if you know you are allergic to honeybee stings, you should carry an EpiPen with you at all times.

Amongst solitary bees, the stinging apparatus of a few, including some *Andrena* species, has, over time, become redundant. Those species that do possess a sting rarely ever use them, and most are so insignificant, they do not have the capacity to pierce human skin. There are just a couple of exceptions. Our tiniest species of ground-nesting solitary bee in Britain and Ireland, of the *Lasioglossum* and *Halictus* genera, both have fully functioning stings capable of penetrating human skin, and can pack quite a punch.

If you put up bee boxes in your garden, or come across one in a public place, you are unlikely to find either of these tiny bees nesting in them, since the boxes are going to attract cavity-nesting bees such as masons or leaf-cutters. Although it is not unheard of for a mason or leaf-cutter to sting, it is extremely unlikely to happen unless you are handling the bees roughly.

Of the other native wild bees we share this planet with, many of them do not even possess stingers. There are the five hundred species in the tribe Meliponini, which live mostly in tropical or subtropical regions. These 'stingless bees' form highly social societies, composed of a queen, some drones, and workers, and collect pollen and store honey in much the same way as European honeybees do.

Only a few stingless bee species store sufficient honey to make them commercially viable for beekeepers, but as they do have some honey stores to protect, you'd have thought nature might have been kind enough to equip them with stingers as honeybees and bumblebees have. In the absence of a sting, these bees usually resort to biting, and their sheer numbers can be sufficient to deter some predators, but by no means all. Some stingless bees have also evolved a 'soldier bee' caste of female worker bees considerably larger than the rest of the colony that are capable of warding off intruders with their bigger bites.

A great many people are more wary of bees than they need be, considering that it is highly unlikely you will be stung in the first place. It seems this fear – which often extends to all flying insects, and anything else with more than four legs – has become endemic in many parts of the world, especially in urban areas where people come into contact with fewer insects on a day-to-day basis than they do in the countryside.

This fear, which is known as *insectophobia* or *entomophobia*, is in many ways irrational (unless, of course, you are allergic to bee stings or insect bites, in which case it is entirely rational). Yet I do not scoff at those who harbour these fears, for despite my absolute love of all things wild, I have to put my hand in the air and say that I, too, have a fear of being stung. Witness my reaction to having a Tree bumblebee entangled in my hair. I have no idea where this fear comes from. It is possible that it was passed down to me by my mother, who was extremely allergic to bee and wasp stings, or maybe it is simply the result of my having been stung a few times in my life, and not particularly wanting to be stung again.

Interestingly, I was bitten some years back by something called a Blandford fly (*Simulium posticatum*), and though I remember seeing it on my leg, and feeling it bite me, I wasn't much bothered at the time, because it was a *fly*. Its appearance held no associations with feelings of pain or discomfort, so I wasn't scared of it. I had no idea how ill it was about to make me. Had I known that being bitten by this fly would result in being bedridden for a week, with a high temperature, flu-like symptoms, and a leg the size of a barrel of beer, I daresay I would have been flapping around like a proverbial whatever the proverbial is. In my case, I think, it must be associations with pain that make me wary of insects that sting. But it doesn't hurt all that much to be stung by a bee or wasp, and bee and wasp stings are not much more itchy than a mosquito bite. So why the fuss and bother?

Many people have simply grown up with a blanket revulsion of insects. I can understand not wanting to get too close to a bees' hive or a hornets' nest for fear of getting stung, and I appreciate that some people have genuine phobias of, for instance, spiders or moths, but I cannot comprehend the instinct to routinely stamp on anything that crawls, slithers, scurries, flutters, or buzzes.

This response – to kill the bug – seems to happen more indoors than out, suggesting that whilst we are able to accept insects in their natural environments, they are less welcome inside our built environments. Maybe this has something to do with being taught that insects are unclean and harbour disease? I don't know, but I do wish more people would try to temper these instincts and habits, and, like David at Berrybrook Farm, cultivate that *live and let live* mindset. At the very least, consider catching insects in jam jars and releasing them outside, rather than killing them, so that they, too, can fulfil their life cycle and do their part in your local ecosystem.

CHAPTER 10

To Bee, or Not to Bee

The sun is shining and our tiny patio garden is alive with flying insects, their combined buzzing, humming, and droning so loud that it all but drowns out the strimmer working the garden next door. I am in heaven – every sense alert, my eyes flitting left, right, up, and down, trying to take it all in – but I am also in purgatory, unable to decide what to watch, or where to rest my gaze. There is so much to take in.

I have already noticed and enjoyed the familiar. Common Carder bees are working the last few blooms on our dwarf comfrey (which I must chop back soon if I want it to flower again this year). Red-tailed bumblebees are on the chives; Buff-tails are bringing pollen back to the nest their queen established deep inside our little rockery this spring. And the very first of this year's Willughby's Leaf-cutter bees is flying back to her nest with pieces of leaf from the enchanter's nightshade that grows along the edges of Sue's tin shed.

So distracted have I been that the lemon verbena tea I made well over an hour ago is now stone cold. I hadn't intended to be outside this long. I take a sip and am pleasantly surprised to discover it is every bit as delicious cold as it is hot. It is such an easy herb to grow, and seems to keep its taste better than others over the winter months after I have dried it. I might make a jug up later and keep it chilled in the fridge. Or perhaps I could try making a cordial with it, as you do with mint or elderflower.

I really should go back inside and get on with my work, but I find myself being enticed by some of the less familiar insects in the garden, those that are not bees. There are metallic bronze shieldbugs (*Troilus luridus*) on the patch of woundwort by the gate. I remember them from last year. And lacewings (Chrysopidae), I think, masquerading as lime-green leaf skeletons. A dainty daytime moth flits around the mint and I wonder if he or she might be a Mint moth. I make a mental note to check my moth book to see if there is such a thing. Each of these insects is extremely beautiful and deserving of my attention, but my gaze settles on a pair of hoverflies. It is their behaviour as much as anything else that pulls me in.

The female is busy foraging on the patch of ox-eye daisies Rob planted in the old rusty wheelbarrow we use as a plant container, whilst the male follows, hovering just above but slightly behind her as she flies from bloom to bloom. He seems to be shadowing her. I assume this must be some kind of courtship ritual. It is quite hypnotic.

Worldwide, there are around six thousand species of the hoverfly family (Syrphidae), with around 280 recorded in Britain and Ireland. Interestingly, this number tallies almost exactly with the number of British bee species, though the bees may soon overtake the hoverflies, as the number of bee species is currently being added to at the fastest rate in a hundred years. I can count on one hand those hoverflies that I am able to confidently identify without Alan Stubbs and Steven Falk's *British Hoverflies: An Illustrated Identification Guide*, and this species, I am happy to say, is one of them. These aerial dancers are Common Drone flies (*Eristalis tenax*), so named because of their close resemblance to honeybee drones. And herein lies a story.

If you were to read a newspaper article about lions, but the accompanying photograph was of a tiger, would you notice? I imagine you would, and you would no doubt be mystified by the muddle. You might even write a letter to the editor pointing out the mistake.

But what if the article were about bees, and it were accompanied by a photograph of a hoverfly? Chances are that, unless you are an entomologist or insect enthusiast, you might not notice. As it happens, the internet and print media are awash with wonderful, well-researched, and extremely interesting articles about bees that have been just so illustrated with photographs of hoverflies. One of the reprinted editions of the world's most well-known reference books on native wild bees, *Bees of the World* by Christopher O'Toole and Anthony Raw, even sported an image of a *fly* on its front cover. The authors must have been dismayed when their wonderfully informative book first hit the shops in this guise.

How could such a mistake have been made? Can hoverfly species really dupe us into thinking they are bees? If so, why do they do it? And how can we everyday folk tell the difference, at a glance, between a bog-standard bee and a big, furry hoverfly?

It was a few years back, when I was still living in Malvern, that I first became aware that some of the bumblebees in my garden were not bumblebees after all. I had spotted something that looked exactly like a Red-tailed bumblebee (*Bombus lapidarius*) foraging on the flowers outside my kitchen window. But there was something unusual about it that I could not quite put my finger on. So I popped outside, took a quick snapshot, and uploaded the image to my computer to take a closer look.

Lo and behold, it was not a bumblebee at all. Although it had been difficult to tell from a distance, I could now see straight away that this insect differed from my beloved Red-tails in a number of ways. For starters, it had large and extremely prominent 'fly' eyes that almost joined together on the top of its head. Bumblebees have ovoid eyes on the sides of their head. Also, this lookalike's antennae were short, stumpy, and hairy. Bumblebee antennae are long and beautifully elegant, and they are set slightly farther apart. On closer examination, I could see that the lookalike was also missing the 'waspish' waist that characterises all bee species. And it appeared to have only one set of wings, rather than two sets, as a bumblebee has.

I was intrigued. From a distance, this creature looked and behaved exactly like a Red-tailed bumblebee, something it was clearly not. It took a bit of research, but I eventually discovered the lookalike's true identity: It was *Volucella bombylans*, also known as a Bumblebee hoverfly. A fly pretending to be a bee. How ingenious.

The Bumblebee hoverfly is a mimic. *Mimicry* differs from *camouflage*. An organism that mimics another seeks to benefit in some way from imitating the behaviour or appearance of the other species, whereas an organism that camouflages itself usually does so by having evolved to blend into the background environment, by having certain colours or patterns on their fur, shell, skin, exoskeleton, et cetera. Camouflaged creatures are usually trying to hide. Mimics are trying to be seen – as something which they are not.

The hoverfly in my garden is playing a particularly special game of mimicry called *Batesian mimicry*. This is where a harm*less* species has evolved to mimic the appearance and warning signals given out by a harm*ful* species. Many hoverfly species have evolved to imitate the colouration and behaviour of bees or wasps. They do this in a cunning attempt to avoid being seen as easy prey by birds and other predators. Some hoverflies are far more convincing than others, at least to the human eye. But humans are not these hoverflies' concern. Those humans who catch creatures to eat, insects or otherwise, usually have time to examine what they are before eating them.

Birds, on the other hand, need to make swifter decisions, especially when they are catching their meals on the wing. To eat, or not to eat? There is often little time to think, so the safest strategy is probably to avoid anything that looks like it might be a stinging insect. This would explain why some of the so-called imperfect or crude hoverfly mimics, which we humans would be very unlikely to mistake for a bee, still seem to gain protection from bird predators whilst looking so fly-like.

Once I realised that such things as hoverfly mimics existed, I started looking out for them in my garden, mentally giving them points out of ten for how closely they resembled the bees or wasps they were imitating. Some were far more like the real thing, known as the *model*, than others. The Blotch-winged hoverfly (*Leucozona lucorum*), for example, is vaguely similar in size and shape to a bumblebee, with striking ivory-white bands across the middle of its thorax, but it's an extremely poor mimic compared with others; I would give it only a 3 out of 10. The Common Drone fly (*E. tenax*), which mimics honeybees, is more impressive – 7 out of 10 for this species, I reckon. To the beautiful ginger-coloured carder bee mimic (*Sericomyia superbiens*), an 8; and to the Hornet hoverfly (*V. zonaria*), a 9. I was going to give the Hornet hoverfly an 8, too, but then I had a chat with Steven Falk, during which we discuss how much more impressive these hoverflies are as mimics when they are on the wing. I am persuaded to up their points to a 9 out of 10. After all, in flight I would not be able to tell one apart from a hornet.

Streets ahead of the others, however, must surely be the Bumblebee hoverfly (*V. bombylans*), which, when I first saw it outside my window, had me completely and utterly fooled. And at a glance, it still does. With its long hair and remarkable bumblebee colourations and markings, this species is an excellent example of Batesian mimicry. Incredibly, it has even managed to evolve a number of different forms, or *variants*, each specifically mimicking a specific bumblebee. The two varieties (abbreviated 'var.') most likely to be found in Britain and Ireland are var. *bombylans*, which we already know resembles the Red-tailed bumblebee, and var. *plumata*, which, with its thick, yellow bands and pure white tail, closely resembles the White-tailed bumblebee and Garden bumblebee. To these bumblebee mimics, I award maximum points of 10 out of 10. They truly are masters of disguise. It is no wonder, then, that so many photo editors mix them up with bees (though I have to say that, in their professional capacity, it might be prudent to check with a scientist before they go to print, rather than just running with some stock image).

There are other bee mimics, too. One of my favourites is the Large bee-fly (*Bombylius major*). The Large bee-fly is one of four *Bombylius* bee-fly species in Britain and Ireland, all extremely good bee mimics. Bee-flies are easily recognised by the way they dart around, often quite close to the ground, hovering occasionally to use their huge, long, extended tongues to collect nectar from flowers such as primrose and speedwell. When they are at rest, it is their wings that give them away. Whilst bees at rest always fold their wings over the back of their bodies, bee-flies hold theirs out at around a forty-five-degree angle from their sides, like the wings on the old Concorde aeroplanes, if you remember them.

Bee-flies can look quite cuddly, resembling tiny, round teddy bears, but their appearance belies the rather cheeky way in which they flick their eggs, which are cleverly laden with grit, into the nests of ground-nesting solitary bees. When their eggs hatch, the bee-fly larvae prey on the larvae of the poor unsuspecting bee. As an aside, the female of the Bumblebee hoverfly similarly lays her eggs in the nests of bumblebees and wasps. Her larvae feed on the nest debris and, occasionally, the resident's larvae, too.

By far the most beautiful bee mimics, however, are the nationally scarce Narrow-bordered Bee Hawk-moth (*Hemaris tityus*) and the Broad-bordered Bee Hawk-moth (*H. fuciformis*). These exquisite creatures have furry, banded, bumblebee-esque bodies and club-shaped antennae, but their most stunning feature is their wings, which are completely transparent, apart from thin veins running through them and the narrow blackish, or broad reddish brown, borders that give each species their respective common names.

Like other hawk-moths, these bee mimics hover, like miniature humming birds, whilst collecting nectar from flowers. I have seen neither of these bee hawk-moths in the flesh, but being eternal optimists, Rob and I have planted their favourite food sources, which, according to Butterfly Conservation, include bugle, ragged robin, wild honeysuckle, devil's-bit scabious, and field scabious. You never know!

Back inside the house, mountains of paperwork are scattered in not-very-neat piles all over the dining room table and floor. They await my attention, but I cannot tear myself away from the garden. I wish, for the umpteenth time, that I had discovered the incredible world of all things invertebrate earlier in my life, because now, there just aren't enough hours in the day to catch up on everything I've missed. But the joy of watching a pair of bee-mimicking hoverflies courting on our ox-eye daisies is not going to get my book written or my filing sorted out.

Flies are primary pollinators for many plants, and play a significant role in the world's crop production. Whilst most are not as hairy as bees, some of the larger hoverflies are extremely hairy, making them effective transporters of pollen. Flies also remain active in unfavourable weather conditions, whilst bees, other than bumblebees, struggle. Flies are often the only insects you will see visiting flowering plants when it is cold, wet, and overcast. They also are more abundant than bees in damp, shady habitats such as woodlands and salt marshes, as well as extreme habitats like those found in Arctic and alpine regions. This means many of the plants that grow in such habitats are more likely to be pollinated by flies than bees. And if you enjoy chocolate, you might be interested to know that the cacao tree, which is incapable of self-fertilisation, is entirely dependent for its pollination on a tiny biting midge. Without tiny biting midges, there would be no chocolate. Imagine the dilemmas this would cause if the cacao tree grew in Scotland!

Thanks to new research, the role of many other non-bee insects is starting to be recognised, as well. However, there are still huge gaps in our knowledge, and as there are rather more flowering plants and floral visitors on the planet than there are scientists to study them, we may never know exactly how effective some pollinators are compared to others. The bottom line is that, to maintain all of Planet Earth's extraordinary, diverse ecosystems, we need to conserve all these pollinating creatures. Maintaining biodiversity is key.

CHAPTER 11

Seeking the Great Yellow Bumblebee, Part 1

I do so enjoy watching nature documentaries. Like millions of other viewers, I am awestruck by the exotic and diverse wildlife brought to the screen, as well as the dramatic landscapes they are filmed in. But for all the brilliance and splendour of the world's rainforests, deserts, and oceans, I still find myself drawn more to the wildlife and wild places of Britain and Ireland than those of faraway places. There is so much I don't know about life in the moors, mountains, fens, forests, rivers, lakes, and shores of these islands we call home, that I barely know where to begin.

Wouldn't it be wonderful if we could see more, hear more, and understand more about the wildlife on our own back doorsteps? Imagine how much richer our lives would be if we made time to lie on the ground and watch life in the undergrowth; to walk through ancient woodlands, meadows, marshes, and heathlands; to immerse ourselves in these different landscapes, absorb their energies, and *encounter*, in the truest sense of the word, other living, non-human beings that live there.

I will never grow bored of the plants and animals living on my own local patch in Dorset, many of which I have still to meet and get to know. I manage my time these days as carefully as I can, to make sure I am free to spend more than just the occasional sunny afternoon watching birds and insects in our garden. And I spend far more time than I probably should looking through the macro lens of my camera at the lichens and mosses growing on Shaftesbury's greenstone walls, and the wild flowers that grow along the tracks and lanes, as I walk between our house and our allotment.

Sometimes, during the months of March and April, Rob and I head out before breakfast in search of boxing hares. We have a few favourite fields where we have seen as many as eleven hares in less than an hour, but I still dream of the day we finally get to see them 'boxing'. Which I know we will, when it is meant to be. And we get up considerably earlier in the month of May to try to catch the dawn chorus at Garston Wood, an RSPB nature reserve just across the border in Wiltshire, near the village of Sixpenny Handley. I say 'try' because we get far too easily distracted by the hares, barn owls, and other wildlife we see along the way, thus frequently arrive late and miss the best of the birdsong. In fact, we sometimes miss it altogether.

All these encounters, and more, I have delighted in, but recently I have felt a yearning to visit other, more far-flung parts of Britain and Ireland that I know are home to treasures I can only dream of seeing near our home in North Dorset. I have, for instance, never seen a mountain hare, a pine martin, or a long-eared owl. There are also butterflies – the Swallowtail, Camberwell Beauty, and Purple Emperor – that I have longed to see since I very first came across their images on the Brooke Bond tea cards I collected as a child.

Then there are the bee species I would love to record, including the Long-horned bee (*Eucera longicornis*), and the Bilberry bumblebee (*Bombus monticola*), which are both famed for their beauty, and the Pantaloon bee (*Dasypoda hirtipes*), who has pollen-collecting apparatus on her legs that make her look like she is wearing jodhpurs. And right at the top of my list, a bee I saw all too briefly a few years ago, on a short trip to the Isle of Barra, but would dearly love to see again – *B. distinguendus*, commonly known as the Great Yellow bumblebee.

This 'distinguished' bumblebee can no longer be found anywhere in England. Because of the data submitted by dedicated recorders, it has been possible to map exactly how diminished the Great Yellow bumblebee's range has become compared with what it used to be. Once widely scattered all over Great Britain, its range has been reduced by 80 percent in the last one hundred years, mostly due, it appears, to changes in the landscape from intensive farming. The bee has completely vanished from England and Wales, and populations are now restricted predominantly to marginal areas of flower-rich, low-intensity crofting in north and west Scotland, the western coast of Ireland, Orkney, Coll, and Tiree, and the islands of the Outer Hebrides.

The islands of the Outer Hebrides. Now, there's a thought. We both fell a little in love with Barra, and Rob has always wanted to visit the Isle of Lewis and Harris.

Though enormously appealing as a destination, the Outer Hebrides are a long way from Dorset. If we were to consider making this journey again, the last thing we would want is for it to be a whistle-stop tour. And spending time away from home might be just the thing we need, for me a welcome sojourn after the sadness of my mother's illness and death; for Rob, a chance to rest his foot, which was badly injured in a fall from a ladder last year and has left him unable to continue working full-time as a gardener. Rob's mother also died last year, so all in all, the last few years have left us both feeling mentally, physically, and emotionally exhausted. They say time is the best healer, and I cannot think of a better way to heal than taking a few months off to immerse ourselves in nature.

So, in mid June, a few weeks after we have laid my beautiful mother to rest, Rob and I pack our cameras, binoculars, waterproofs, bicycles, sunscreen, midge deterrent, maps, and guides to the flora and fauna of Britain and Ireland and head north in our camper van to the Outer Hebrides. Our plan is to spend the entire summer exploring these islands on the edge of the Atlantic. But we both hope that, whilst we are there, we might happen, again, upon Great Yellow bumblebees.

Our trip could not be more perfectly timed, as it coincides with the flowering of the *machair*, a rare and beautiful costal habitat found only in the north-west of Britain and Ireland. Machair is Great Yellow bumblebee habitat. Although the machair flowers from May through to August, you are not likely to see Great Yellows, which are late-emerging bees, until late May or early June, and you won't see them in abundance until well into

August. If luck is on our side, we should be on the islands long enough to see queens, recently emerged from hibernation, some July workers, and, if we don't run out of money, a new generation of August queens and males.

As it turns out, our journey, and the creatures and plants we meet en route, are every bit as memorable as our final destination.

We leave Dorset on a Thursday, hoping to arrive sometime by the end of that weekend on the northernmost island, Lewis and Harris. Thereafter, we have no fixed plans, other than to slowly work our way south towards the Isle of Barra, spending our days walking, cycling, and basking in (or braving) the Hebridean weather. But plans are made to be changed, and less than forty-eight hours after we set off, we find ourselves standing, not on the quayside at Ullapool, but by the side of a small loch on the edge of the great Caledonian Forest of Scotland.

We were supposed to be catching tomorrow morning's 10.30 ferry from Ullapool to Stornaway, but there are 'weather warnings'. I am prone a little to feeling seasick, and a lot to worrying about boats sinking, even on seas that others tell me are as 'still as millponds', so the term 'weather warning' terrifies me. Fortuitously, just moments after hearing the weather report on the radio, I spot a road sign that tells us we were not far from RSPB Loch Garten, a world-famous osprey breeding site. This gives me the perfect excuse to suggest taking a detour in the hope that we might see one of these magnificent birds of prey whilst sitting out the upcoming storm.

Osprey (*Pandion haliaetus*) used to be found throughout Europe, but due to sustained persecution, combined with the Victorian vogue for taxidermy as well as collecting birds' eggs, they became extinct in Scotland in 1916. However, in 1954 a pair of breeding birds, believed to be of Scandinavian origin, returned to Scotland and set up a nest. Since 1959, osprey have been breeding successfully at Loch Garten. The RSPB set up 'Operation Osprey' to protect these birds from ongoing threats, including modern-day egg thieves who continue to target and plunder their nests. Without the tireless efforts of conservationists, volunteers, and landowners, the ospreys might not have been so successful.

Excited by the prospect of seeing osprey, but arriving late in the day, just before the visitor centre is due to close, we pretty much abandon the van in the car park. I grab my camera and a bottle of Avon Skin So Soft – apparently the only truly effective defence against the dreaded Scottish midges – and we head straight for the centre.

The staff, mostly volunteers, are extremely helpful and keen to tell us about the wildlife on their reserve. They call our attention to three siskins on the feeders just outside one of the observatory windows; there are chaffinch, too, and a greater spotted woodpecker, but he is shy and flies off when a pair of pigeons land clumsily on the branch next to him. The osprey nest is visible in the distance through window slots in the sides of the building, and there are binoculars and monoscopes for our use. It is far easier to see the nest on the centre's live video feed, but I feel more connected looking through binoculars. I can see a clutch of three eggs, but no sign of the adults.

The volunteers inside the visitor centre explain that this year, for the second season in a row, the twenty-year-old female known as EJ has failed to raise any chicks. She has been breeding here for fifteen years, and in that time has successfully raised twenty-five chicks. However, her new mate, George, has not provided her with sufficient fish this year, and she has been forced to abandon her eggs and go in search of food for herself. This is not something she would have done lightly, we learn, as this particular bird continued to incubate her clutch the previous year despite, at one stage, being completely buried in snow. George, it seems, is young and inexperienced – unlike EJ's previous mate, Odin, with whom she has successfully raised seventeen chicks in the last nine years.

The staff are clearly saddened by the current situation, but tell us this is infinitely preferable to what happened the year before. Then, it is thought that, a few days after the chicks hatched, their father Odin was chased off by a rival male, and the chicks all starved. It must have been heartbreaking for the staff to watch this play out on the live webcam. I want to ask them why they didn't feed the chicks themselves but resist, as I already know the answer, which is that conservationists do not intervene in such situations, instead leaving nature to take its course.

After the centre closes up for the evening, we return to our camper van. The wind is picking up, so we hunker down in a small clearing just off the road. The view is spectacular. A lone goat willow (*Salix caprea*) frames the window to our left, whilst ancient Scots pine (*Pinus sylvestris*) grow all the way down to the edge of the loch on the right, some of the branches stooping down so low that they almost touch the water. Between the trees, and directly in front of us, is a narrow beach, no more than two metres deep and fifteen metres long. On the far side of the loch, and behind us, more pine, as far as the eye can see, and beyond. We are in the Abernethy Forest, the largest surviving remnant of the Caledonian Forest that colonised Scotland at the end of the last ice age.

A smallish bird flies in from the right, plops down in the surf at the water's edge, and begins to work his way along the beach from the tall dark pines to the old willow. He has come to feed. Pickings must be rich, because once he has reached the goat willow, he flies all the way back to the pine bank at the far end of the beach and starts over again, his rear end bobbing up and down as he struts along the surf line.

He reminds me of a wagtail, but with a much shorter tail, and his legs are proportionally much longer, too. Now that I am paying closer attention, I see he is not so much bobbing as 'hiccuping', and I don't think wagtails do that. I have seen this bird before and know I should be able to name him, but cannot. 'Common sandpiper,' Rob tells me.

I wish I knew my birds better and that I had not wasted so many precious years away from all this. But I catch myself on, for life is short and there is no time for regrets. I banish the what-ifs and bring myself back to the here and now. The weather outside is wild and I am sitting at the edge of a loch in the Scottish Highlands, protected from the worst of the elements in our old but comfortable van, surrounded by soaring Scots pines and the wildlife they are home to. Somewhere out there, unseen but seeing, are red squirrels, pine martins, otters, crested tits, and osprey. I count my blessings and give thanks to the universe for the storm that caused us to take this detour.

When we wake the next morning, there is time for a walk before we continue on our journey. The area around the van is quite boggy, even more so after last night's storm, so we decide to head away from the loch and up into the forest. It is thanks to its bogginess and inaccessibility that this area of the Caledonian Forest, around Loch Garten, is still standing. Had it been easier to access, it, too, would have probably been destroyed by man.

Scots pine is the forest's largest and longest-lived tree. It can live for up to seven hundred years and is an important *keystone species*, a species on which an entire wildlife ecosystem depends. It is incredible to think that many of the mighty trees we are walking beneath today were around when wolves still roamed here, and beavers, too. Further back in time, this ancient forest, which once covered 2.7 million acres, was also home to lynx, wild boar, and bear.

There are no top predators left in Britain and Ireland now, but anything is possible, and who knows what the future might hold, given some of the forward-looking rewilding schemes being discussed. Already, thanks to the Scottish conservation charity called Trees for Life, the ancient Caledonian Forest and its existing wildlife are slowly, but surely, being brought back from the brink. With the help of volunteers and conservationists from around the world, the charity has planted well over a million trees, including birch, aspen, willow, oak, and elder as well as Scots pine, and the natural regeneration of several hundred thousand more is well under way.

Scots pine are, according to Trees for Life, 'unusual amongst conifers in having a number of different growth forms, ranging from tall and straight-trunked with few side branches, to broad, spreading trees with multiple trunks'. When they grow close together, like these, they possess very little in the way of branches or foliage until fairly close to their tops, so as we walk, we are mostly surrounded by trunks. The barks are thick and shallowly fissured, grey-brown at the bottom but orange-brown towards the top. The orange is glorious in the early morning light.

We climb over a fallen tree, its tangled roots splayed like a giant cartwheel above the empty crater they once grew in. The roots are full of nooks and crannies for small birds and mammals to hide and nest in. This wood is rich in resins, which means it decomposes more slowly than other woods. It is soft, pitted with crevices and holes, and encrusted with mosses, liverworts, and lichen – a treasure trove for insects which, once I get my eye in, are too numerous to count. There are also some trees that are clearly dead, but still standing. I heard recently on a radio programme that these pines can stand, dead, for up to one hundred years. Whether fallen or standing, all trees support even more species in death than they do in life.

We are just about to move on when I notice a patch of shrubby plants growing beside the fallen tree. They have small waxy leaves and clusters of pretty, little white bell flowers. I wonder if this might be cowberry? We have left the wild flower book in the van, so I take a photograph of the flowers, and one of another plant with long, thin leaves and pale-yellow trumpet-shaped flowers that I don't recognise at all. These are the first of many new plants we are likely to encounter as we travel north. I can't wait to find more. We walk on, deeper into the forest, and I am surprised by how light it is. When I turn to look back, the absence of understory gives me a clear view all the way down to the water's edge and across the loch to the other side.

The ground beneath us is now quite sandy, strewn with pine needles, scattered broken twigs, and pine cones. We stuff some of the cones into our pockets. They will come in handy when we make a campfire – not here, of course, but perhaps on a big empty beach when we reach the islands. We scan the canopy for signs of movement, hoping against hope that we might catch sight of a pine martin, a red squirrel, or perhaps a flock of crested tits or Scottish crossbills. But if there is anything up, there it remains invisible to us; the Caledonian Forest is not giving up its secrets today. Maybe another time.

We head back to the van, where we bump into a walker who tells us there is a red squirrel feeding station just up the road and that we are more or less guaranteed to see one there. This is too tempting for me to dismiss. It's not quite the same as catching a glimpse of one of these beautiful and iconic creatures scampering up a tree in the middle of the forest, but it's a darn sight better than not seeing one at all.

We do see red squirrels at the feeding station, two of them, and they are every bit as delightful, and red, as I expect them to be. Because these particular individuals are not afraid of humans, they allow us to come quite close; one is bolder and completely unfazed by our presence, and sits for a while on a branch right above the feeding station. I have my camera to hand, so watch it through the lens.

Its hind feet are spread flat on the branch, which it grips tightly with ten sharp little toenails, whilst its front feet – *or should I call them paws?* – clasp a nut. The squirrel's head is bent forward as it nibbles the nut, and its tail follows the curves and contours of its body all the way up to its neck. It has beady, black eyes and cream-coloured tufts at the tips of its ears – the spitting image of Squirrel Nutkin. I am so sad that this beautiful creature and its kind now need special conservation measures to ensure their survival.

I am aware that, although I feel sad about the conservation status of red squirrels, I am not experiencing any kind of 'connection', deep or otherwise, with this individual. I am simply observing it, much as one might observe an animal in a zoo. Seeing it in its natural environment makes the experience more personal, but at the end of the day, I am basically taking advantage of the fact that it has been partially tamed by the easy availability of food in the feeding station. It doesn't really feel like a 'wild' encounter to me.

I wonder at what stage an encounter with a wild animal becomes something more than merely an 'encounter,' an unexpected but fleeting meeting. This makes me think about the birds that come to our feeding station at home. Some, including the robins, blackbirds, and long-tailed tits, appear to have become tamer as they have come to trust us, and I am touched to feel this trust.

Others, like woodpeckers and bullfinches, remain as wild as wild can be. I would love to experience the sort of longer-lasting, permanent connection with wild creatures that is known by those humans who, living gently alongside them in remote and wild places, are fully accepted as just another part of their natural surroundings, much in the way you sometimes see rabbits comfortably nibbling grass alongside birds, themselves digging for worms. In these situations, although there is no exchange or interaction between species, there is nevertheless trust and acceptance.

———

We spend our final night on the mainland in a lay-by above the port of Ullapool. We like Ullapool. It feels like a meeting place, a mixture of old and new, with lots of comings and goings but somehow still grounded in community. The locals probably think otherwise, but despite the many tourists, it doesn't feel to us like a tourist town.

The sun is shining and we have a few hours to kill, so we have coffee and lemon polenta cake on the seafront. Rob gets lost in a proper, old-fashioned hardware store; right next door I find a wonderful independent bookshop, where I buy a copy of John Lister-Kaye's *Gods of the Morning: A Bird's Eye View of a Highland Year*.

Finally it is time to board the ferry. Storm Hector has been and gone and our journey to the Outer Hebrides is about to begin.

The ferry crossing takes around three hours and we spend most of it on deck watching large groups of gannets (*Morus*) dive-bombing for fish. Rob has seen this many times before, but I haven't, and I find the whole spectacle – for a spectacle it certainly is – thrilling. The gannets circle in crowds, high above the sea, searching for fish. They must have extraordinary eyesight. When I look over the side of the boat, I can see absolutely nothing beneath the turbulent, grey surface of the water; the gannets, on the other hand, see fish.

Once they have a target in sight, they peel off and plunge dive at speeds of up to one hundred kilometres per hour, folding their wings back closer and closer to their bodies until they morph, just milliseconds before they hit the water, into living torpedoes. The timing and accuracy as they pierce the surface are stunning, as is the fact they don't break their necks in the process, or collide with one another either above or below the water's surface. They must possess extraordinary spatial awareness. I have since read that gannets have specially developed neck muscles and a spongy bone plate at the base of their bill to reduce the impact as they hit the water.

Once the gannets are beneath the water, the momentum the birds have built up during their sky dive helps propel them to a depth of up to five metres, after which they have the ability to swim, or 'fly', underwater, using their wings much in the same way as they do in the air. They often swallow the fish they catch whilst still beneath the surface. This reduces their chance of being harassed by great skuas (or 'bonxies', as they are known in these waters).

I don't blame them. Fancy going to all that trouble only to have your hard-earned catch stolen by a marauding, bully bird, whose aerial acrobatic skills are so finely tuned that it is able to grab a fish from another bird's beak in mid-air and never has to fish for a meal of its own. Interestingly, the word 'gannet' is derived from Old English *ganot*, meaning 'strong or masculine'. That seems far more appropriate than the way we now use this word, in a rather derogatory way, to describe people who are greedy eaters. We ought to call those people bonxies.

Gannets are distinctive-looking birds, ungainly but beautiful on land and utterly magnificent in flight. Their expressions are slightly dour, almost disapproving; their eyes and beaks, which appear to have been attached to their faces and then defined around the edges in black, felt-tip marker, remind me of medieval plague doctors, who wore masks with birdlike beaks packed with dried flowers, herbs, and spices to protect them from becoming infected by their patients. With a wingspan of around two metres, gannets are easily Britain's largest seabird. I have read they can live for more than thirty years, first breeding when they are four or five years old and staying with this same mate for life, in colonies of up to 150,000 of their kind.

This calls to mind the tragic story of Nigel the gannet, which made the news several years ago. Nigel arrived on the small island of Mana, off the coast of New Zealand's North Island, in 2013, having been lured by the sound of gannet calls. Alas, these calls were not made by real gannets; they were being broadcast by solar-powered speakers that had been placed on the eastern cliffs of the island, alongside eighty concrete decoys with painted yellow beaks and black-tipped wings, to attract a colony to settle there. Only Nigel came.

Nigel soon picked out a mate from amongst the concrete birds and, according to conservation ranger Chris Bell, who lives and works on the island, carefully constructed a nest for his mate from seaweed and sticks. He groomed her cold, concrete feathers and chatted to her 'year after year after year'. Nigel remained on the island, faithful to his mate, until his death in early 2018, just a few weeks after three real-life gannets finally arrived on Mana Island. Nigel sadly failed to befriend the new arrivals, preferring the company of his concrete colony. Bell was devastated by Nigel's untimely demise, but as gannets like to nest where other gannets have nested before, it is believed his presence on the island played a major part in attracting the other birds. 'It's really sad he died,' Bell told reporters, 'but it wasn't for nothing.'

We are bound to see more gannets, Rob tells me. In the Outer Hebrides, they are not in short supply. And so I turn my back on the noisy, torpedoing gannets that are fast becoming specks in the distance and look ahead, with great anticipation, as we approach our next port of call, Stornaway.

I had thought, whilst we were planning this trip, that our 'adventure' would begin when we landed on the islands, but in reality the whole journey is turning out to be an adventure, and being surrounded by gannets on our crossing has undoubtedly been one of the highlights so far, especially for Rob, who has a soft spot for these seabirds. It is wonderful that, through knowing Rob, I have come to share his interest in birds, as he has come to share my interest in wild bees. These shared interests make our lives together even more meaningful and fun.

I feel a great sense of anticipation as we get ready to disembark, tinged with sadness that I will not be able to share tales from any of our upcoming adventures with my mother, who would have loved to hear them. I hope this sadness will soon be soothed by the wonder and wildness of the Outer Hebrides, and the fact that we have no particular plans or agenda for the next few months, other than to go, and be, wherever our fancy takes us.

CHAPTER 12

Seeking the Great Yellow Bumblebee, Part 2

It is overcast when we arrive on Lewis, and my first impression as we leave Stornaway is that this is as bleak and barren a place as I have ever visited. For miles and miles, all we see is an apparently empty landscape and wide open skies. But this landscape is not so desolate as it first appears.

The first thing that catches my eye is the cotton grass, difficult to miss as its fluffy, white cotton tufts bend and sway in the wind on top of their long, sedgy stalks. I love this plant, even when it is wet and waterlogged. There are grasses, too, and spongy sphagnums, and the surface water is brown. Look closer, in and around the shallow ditches and gullies that are a feature of this peaty moorland, and you find clumps of lousewort and bog asphodel. The lousewort is pink and flowers on short stems with stunted foliage. The bog asphodel is taller than the lousewort, but shorter than the cotton grass, with clusters of reddish buds that open into pretty, yellow star-shaped flowers. I am surprised by how delicate these little flowers are.

They must be sturdier than they appear, for you need to be seriously sturdy and resilient to grow in this landscape, or otherwise you would not survive.

Another group of plants that have adapted to cope with these elements, and with the acidic, waterlogged conditions of the peat bogs, is sundew (*Drosera* spp.). I have not seen this plant or anything like it before, and it intrigues me. Sundews are small and grow very close to the ground, so you need to look at them through a magnifying lens to really appreciate their beauty. Their leaves, which grow out from the roots in a rosette, are a yellowy green and remind me of flattened thumbs. I think these are 'oblong-leaved' sundews (*D. intermedia*). Each leaf is edged with short, spiky red hairs, a bit like tentacles, and at the end of every hair is a single droplet of dewy liquid that shimmers and glistens in the sunlight, hence their name. This plant is carnivorous, and the sticky substance secreted on the hairs acts as a lethal trap to any unsuspecting insect that stops to take a sip. Some of these sundews have produced tiny white flowers on long, straight stems that stand high above the level of the leaves. I guess the flowers need to be well out of the way of the sticky secretions, as it certainly wouldn't do to trap and digest the insects you need for pollination. I wonder which insects actually pollinate this plant. I don't see any bees around; flies, perhaps.

There are heathers, too, everywhere, as far as the eye can see. I hadn't noticed them at first, because they are not yet blooming. It won't be long before they burst into colour, and when they do, they will paint the peatland purple. It is from the heathers that the Isle of Lewis gets one of its Gaelic names – *Eilean an Fhraoich*, meaning the 'Heather Isle'.

I am beginning to see this vast open landscape in an entirely different light now. Far from being barren, this unique habitat is teeming with life. Not only does it support a diverse community of plant life and invertebrates, but it is also an internationally important habitat for breeding birds, including dunlin, golden plover, and curlew. Hen harriers and short-eared owls depend on this habitat as a hunting ground, too.

Most important, such peatlands play a vital role in locking in and storing carbon. It is more vital than ever before that habitats like this one on the Isle of Lewis are protected. Whilst they are in good condition, peatlands outperform any other ecosystem as carbon sinks. Of the entire world's surface, only 3 percent is peatland, but this meagre 3 percent stores at least twice as much as the carbon stored by all the forests standing on the planet.

According to the Woodland Trust, 'If peat bogs are damaged or degraded, they begin to emit carbon dioxide and other greenhouse gases. A loss of just 1.5 percent of the world's peatlands releases the equivalent of all the carbon emissions humans create worldwide in a year.' If you are someone who still buys peat-based compost for your garden, perhaps now might be the time to switch to something more sustainable.

So engrossed have I been in meeting these new plants that I have completely lost track of the time. We need to find somewhere to park up, so we leave the moors and peat bogs and travel up to the northernmost tip of the island to spend the rest of the evening by the harbour at Port of Ness. There is a flock of gannets fishing in the bay tonight. They are so close, we don't need binoculars. Behind us, fulmars are nesting in old rabbit holes on the cliffs. A tern, with a sand eel clasped in its tiny beak, battles against the might of the wind as she presumably makes her way back to her nest, which lies somewhere inshore, hidden in the long grass beyond the cliffs. The beach is empty, the weather is wild, and there is still daylight at midnight.

From Port Ness, we travel south, via the Callanish Stones, which pre-date Stonehenge by around two thousand years. Here we celebrate the midsummer solstice. I love to mark the changing seasons, and feel sad that the solstices have been hijacked by some who take them as an excuse for a giant party. It is not the party itself I object to, but a lack of sensitivity and respect for the landscape and ancient monuments around which they gather, and which too often they have trashed before the sun has even had a chance to rise on this, the longest day of the year. Thankfully, the people we meet at Callanish care deeply about the stones and the land they stand in, and so it happens that we end up watching the sun rise on what must be the coldest, but by far the most beautiful and magical midsummer dawn I have ever experienced.

We have been on these islands for less than a week, yet already my senses are overloaded and my brain can barely keep up with all the sights and information it is trying to process. I have been keeping a nature diary since we arrived, noting down the name of every single plant, bird, and insect we see, and I am beginning to feel like a child let loose in a sweet shop. I cannot believe how many different flowering plants I have noted – at least thirty-five thus far, and that's just the ones I have been able to identify. The others I hope to identify from my photographs when we get home in August.

We have also seen huge numbers of birds, not just seabirds and waders, but also rock and meadow pipits, skylarks, and wheatears, so many wheatears that I think half the world's population of this species must surely be visiting the Isle of Lewis. I have become quite fond of these funny little birds. Less timid than others, they hop and flit alongside us, keeping just a couple of fence posts ahead, as we cycle or walk along the island's narrow roads. They are similar to robins in size and shape, but more striking. The males have blue-grey heads and backs, black wings, black cheeks, white stripes above the eyes, and pale orange-flushed chests and necks. The females are brownish and less easy for me to identify. I get them confused with other species.

We have heard the calls of cuckoos, skylarks, oystercatchers, lapwings, curlews, and a corncrake (*Crex crex*). Yes, a corncrake – which, I swear, was only metres from the van. So well hidden was he in the undergrowth, that despite our extreme patience, over a number of very frustrating hours, we still failed to catch sight of him. We didn't even see the grasses move, though we know from following his constant *crex-crex*ing that he was continually on the move. They don't call these birds 'elusive' for nothing.

Most thrilling of all our sightings, so far, has to be a close encounter with a pair of white-tailed eagles (*Haliaeetus albicilla*), that we happened upon on the morning we left Callanish. These eagles were not out at sea but inland, cartwheeling through the air straight at us, claws clasped together, just a few metres ahead of the van. It looked at one stage as though they were going to hit the ground, but they separated from each other in the nick of time and veered off, up, and away. We followed them through our binoculars until they disappeared from sight. There are no superlatives to describe these birds, or how they made me feel. I will never forget the experience.

And still we have almost two full months ahead of us. No deadlines or commitments, and no agenda other than to absorb the sights, sounds, scents, and energies of these islands. I wonder what it would be like to experience the Outer Hebrides in autumn, winter, and spring. Perhaps we shall one day. Perhaps, tomorrow, we will find a deserted croft on the edge of the machair, somewhere between the mountains and the vast Atlantic Ocean, where we can park up our van, become caretakers of a little plot of land, and never have to leave. I wonder how many who have visited before us have had exactly the same idea.

What is it about being somewhere wild that causes such longings to surface? There is something primeval in how we yearn to feel the sand beneath our feet, swim in the sea, hug trees, dance under the stars, and walk to the top of the highest mountains to experience the weather in all its might. But not everyone responds to wild landscapes this way. For such abandon to happen, you may need already to be consciously *open*, that is, ready to *allow* or *will* oneself to connect physically and emotionally with the natural world. Maybe those who don't feel this way have developed a resistance to the tugs and senses I am experiencing right now. Maybe they are frightened. Maybe they are worried that if they allow their barriers and shields to temporarily be disabled, and expose themselves to some primal calling to be at one with the elements and the Earth, they might never want to go back again.

Fear is a funny thing. Sometimes the strongest people you know are more frightened of 'getting in touch with their more sensitive side' (as the saying goes) than they would be of facing an attack from a lion or tiger. I cannot help thinking that if the world's politicians and business leaders were to spend more time in wild places like this, leaving behind their shoes, socks, job titles, and mobile phones and immersing themselves wholly in nature, they might just catch a glimpse of something bigger and more important than the 'economic growth' they seem so obsessed with. They might realise that you cannot put a price on nature, and that prioritising the economy over ecology is not only short-sighted but also, in the long term, suicidal. I wish.

———

Ever since a brief visit to the Isle of Barra a few years ago, Rob and I have both longed to come back. The last time we were here, we caught a glimpse of Great Yellow bumblebees, but our time was cut short. This trip we are hoping to have a chance to really observe them. We have also noted that these islands are home to Moss Carder bumblebees (including the rarer Scottish island form of this bee which is sometimes called the 'Hebrides bumblebee'), another species whose numbers are dwindling, as well as a very rare solitary bee, the Northern Colletes bee (*Colletes floralis*). These islands are a seventh heaven for insects, birds, wild flowers – and me.

Lewis soon becomes Harris, and the landscape changes again. We travel south on the Golden Road, a narrow single track with passing places that twists and winds its way through the dramatic eastern coast of the Isle of Harris. The terrain is quite extraordinary, encompassing mountains, moonscapes, and countless inland lochs, inlets, and bays. Between the inlets and bays, growing on extended areas of low, flat ground, there are swathes and swathes of bright pink thrift.

I am struggling now to keep up with my nature diary. So varied is this landscape, and so abundant the wild flowers and grasses, that each day brings with it a plethora of new scenery and species, and now that we have crossed to the west coast, the landscape takes my breath away. This was never going to be a beach holiday for us, but it is difficult to do anything other than sit, open-mouthed and wide-eyed, as you gaze upon these pristine white sands and azure waters. It is here, just beyond the beaches, that we get our first real introduction to the machair.

It is spectacular. Wild flowers, more species than I can count, thrive and bloom in this rare and fragile coastal habitat. Machair comprises different environments – some wet, some dry, and some peaty, but predominantly coastal and sandy – so depending upon the location and time of year you visit, the plants you see will vary.

Picture, if you can, a wild flower–rich landscape where wild orchids bloom in abundance alongside milkwort, eyebright, and wild thyme; where ragged robin, cotton grass, and buttercups sway, on their long, leggy stems, in the breeze above an understory of clovers and vetches; and where the air is filled with the sweet, gentle perfume of lady's bedstraw. Self-heal, knapweed, harebell, and forget-me-not grow amongst them; so, too, do plants with unfamiliar but mouth-wateringly delicious names such as sneezewort, procumbent pearlwort, and amphibious bistort. The list goes on and on.

When you are ready to tear yourself away from individual flowers, try arranging them by colour or hue. The machair landscape shifts from predominantly white in May to every conceivable tone of yellow in June, through riots of midsummer pinks, reds, and mauves to late August blues and purples. Add to this the backdrop of white sand and ever-changing seas… or face inland to a contrasting scene of mountains dotted with the blue-brown waters of peaty, inland lochs, each with their own tiny islands waiting to be explored. If you can see any of this in your mind's eye, you might begin to conjure up an impression of this stunning landscape. If you are ever fortunate enough to visit it in person, I promise it won't disappoint.

The Gaelic word *machair* means 'fertile plain', but it is a combination of coexisting variables that creates these unique areas between the sea and the peatlands. This mix includes low-lying grasslands; sand, enriched with the fragments of millions of shells and held together by deep-rooted marram grass; the effects of strong onshore winds blowing the sands inland; the right amount of rainfall; and, most crucially, the involvement of people and their grazing animals. 'So unusual is the right combination of these features', say Scottish National Heritage, 'that machair is restricted worldwide to just the north-west of Scotland and the north-west of Ireland.'

For more than a thousand years, crofters have managed long narrow strips of machair using the low-impact, traditional practices of seasonal grazing and crop rotation. Whatever is left after cropping is ploughed back into the soil, which stirs up the seed bank and allows wild flowers to grow in amongst the crops. Only seaweed is used to fertilise the land. As no herbicides are applied, the wild flowers flourish.

For seven glorious days, Rob and I explore and absorb the beaches and the machair of western Harris, and it is here, in a patch of clover, that we find our first Moss Carder bumblebee (*Bombus muscorum*). Like the Great Yellow bumblebee, this species has undergone steep declines as a result of changes in agriculture, and possibly climate change, over the last century, though it is still more widely distributed across Britain and Ireland than the Great Yellow, with a number of coastal strongholds in Wales, England, and Scotland, as well as here on the Outer Hebrides.

Interestingly, the Moss Carder bumblebee's floral preferences are similar to those of the Great Yellow. This suggests that other drivers might be at play when considering why the Great Yellow bumblebee's range has been squeezed so very far north, whilst the Moss Carder bumblebee still thrives in more southern parts of the United Kingdom.

It is with heavy hearts that we leave Harris, for we must catch a ferry, so once we leave we cannot easily change our minds and come back. But it is time to move on now, via Berneray, to the Uists.

By the time we reach RSPB Balranald, a nature reserve on the western coast of North Uist, it is early July. Our plan is to park up here for the better part of a week and use our bicycles to get around. The reserve is well known for its sweeping sandy beaches, rocky foreshore, marshes, dunes, and machair. Here, we hope to see large numbers of seabirds and waders, and, if we are lucky, actually lay eyes on a corncrake. I am also desperate to find Great Yellow bumblebees. But first we need to unload and set up camp.

Whilst we are levelling the van, I become aware of a group of people with insect nets and camera equipment in the field just beyond the campsite. The field they are standing in, and pretty much all the other strips of land surrounding it, are covered in yellow flowers. I cannot tell for sure, but I think they might be kidney vetch (*Anthyllis vulneraria*). Great Yellow bumblebees like kidney vetch. I wonder what the people are doing, and notice they are mostly looking down at the ground. There are organised wildlife tours on the Outer Hebrides, so I imagine this group must be looking for butterflies or bumblebees.

I am becoming impatient watching them. 'Go!' says Rob. 'I'll follow on when I've finished unloading. And don't forget your camera.'

I abandon the van, the bicycles, the unloading, and Rob, as I half walk, half run, towards the field, as the people with nets and cameras are beginning to walk away. I want to catch them to ask if they have, by any chance, seen Great Yellow bumblebees. Fortunately, they are walking in my direction, and I meet them at the entrance to the campsite. I can see now, from the equipment they are carrying, that they are a TV crew. This is promising.

'Forgive me for intruding,' I start, 'but I was watching you in the field over there. Were you by any chance looking for Great Yellow bumblebees?'

The answer is affirmative. In fact, it was Great Yellow bumblebees, specifically, that they were looking for and recording, for a slot in a two-part special of BBC's weeknight programme *The One Show*. They will be broadcasting live from the Outer Hebrides the following week. From what they say, there are quite a few Great Yellows still foraging on the kidney vetch right now.

I wave frantically at Rob, beckoning him to come quickly. I don't want to miss the bees, but I also want him to be there when I find them. I wonder if they will be as beautiful, and as yellow, as I remember them being.

We pick our way with great care through the carpet of kidney vetch to the spot where the TV crew were filming. There is nothing here now, so we walk back along the edges of the kidney vetch, eyes down, stopping every now and then to listen. With bumblebees, it is often the deep reverberating hum that alerts you to their presence, and sure enough, within just a few minutes, I hear one.

I tune in, turning my head so I can home in better on her position. Suddenly, I see her – a beautiful, pristine Great Yellow bumblebee queen. She is every bit as 'distinguished' as her name, *B. distinguendus*, suggests she should be. She is clothed from top to tail in a thickly piled, bright ochre coat, interrupted only by a single bold, black band right across the centre of her back, between her wings; her underside and legs are entirely black. This bee truly is a queen amongst queens.

My very distinguished bumblebee queen happens to have a strikingly deep buzz, making it easy to follow her as she flies purposefully from flower to flower. With her long tongue she is probing the deep flower heads on the kidney vetch for nectar, and I can see she is already carrying pollen on her hind legs. This suggests she has begun to establish a nest, and is probably collecting the pollen to provide for her first brood.

Back on the mainland, some bumblebee species are coming to the ends of their nesting cycles, but because Great Yellows do not emerge from hibernation until late May or early June, when their preferred flowers are blooming, they are late starters.

Like many other bumblebee species, Great Yellows have a preference for nesting in the abandoned nests of rodents and rabbits. There are a multitude of these hidden beneath the tussocky grass of the machair. However, although this habitat affords perfect conditions for nesting, I have read on the Bumblebee Conservation Trust website that nest density has been found to be 'no more than one or two nests for every square kilometre of suitable habitat'. Also, this species produces small colonies compared with other bumblebees. Given its nesting biology, and the lack of suitable habitat remaining in Britain and Ireland, it is no wonder this bee is nationally scarce.

I am thrilled to have found a Great Yellow bumblebee so soon after our arrival at Balranald, and feel optimistic that we will find more. This is clearly their kind of habitat; they have everything they need to successfully complete their life cycle here. As well as the kidney vetch, there are other favourites, such as bird's-foot trefoil and red clover, and it looks as though there will be a bumper crop of knapweed when it comes into flower later this summer, which will sustain the bees after the vetches, trefoils, and clovers have all gone over.

It is vital that bumblebees have suitable flowers to forage on throughout their life cycles, but habitats like these are now sadly few and far between. We have lost around 98 percent of our wild flower–rich grasslands in the United Kingdom since the end of the Second World War, and alongside these losses we have seen declines in not only Great Yellow bumblebees and other wild flower–loving species of bees and butterflies but also ground-nesting birds such as lapwing, skylark, curlew, and corncrake, all of which like to nest in wild flower meadows and grasslands.

We do indeed find more Great Yellow bumblebees, almost exclusively on the kidney vetch. But they are nowhere near as numerous here as the Moss Carder bumblebees, especially on days when the weather is more overcast. This gives me cause to wonder, again, whether there might be other factors at play in their decline.

An online chat with a friend prompts me to dig a little deeper, and I find a paper written by Dr Paul Williams from the Natural History Museum in London. Williams suggests that Great Yellow bumblebees might be more vulnerable to climatic changes than are other bumblebee species. Where species decline is concerned, things are never as simple as they seem. There are multiple drivers, and it is really important not to lay the blame at just one door. It is too easy to point fingers at just one factor, for instance, pesticides, or the rise in diseases and invasive species, but in actual fact, the causes are complex and they need to be considered and tackled holistically, not in isolation.

Our week on North Uist soon comes to an end. The number of flowering plants I have seen and identified is now approaching seventy, and our bird sightings on this isle include three short-eared owls, a golden eagle, and numerous hen harriers, both male and female. The Outer Hebrides are a safe haven for these and many other raptors, many of which I am unlikely ever to see back home in Dorset.

I feel especially privileged to have seen hen harriers, and deeply saddened to know they are so mercilessly persecuted on the mainland. I hope, one day, that they will receive the protection they deserve and that their populations will increase throughout Britain and Ireland, and not just on the remote islands of the Outer Hebrides.

The cherry on the icing on the cake comes the day before we leave, when we are unexpectedly treated to a sighting of *two* corncrakes. Although we have been hearing them *crex-crex*ing day and night, we had completely given up on looking for them, and so were all the more startled when a pair flew up from the ground, flapping clumsily, right in front of us. They crossed the field and dropped down over the far fence, never to be seen, at least by us, again.

———

On the morning of our departure, Rob looks at the map. He enjoys looking at maps. I'm glad he does; otherwise, we might have missed several of the most stunning areas of machair we have visited.

'Do you fancy going to Baleshare today?' he says. 'Apparently, it's a birdwatching hotspot.'

'Why not,' I reply.

And so, a few hours later we cross the causeway that leads to the tidal island of Baleshare, off the west coast of North Uist. It is extremely flat. We pull up just above a beach, and I walk back along the road to an area of machair that caught my eye from the van as we passed it.

Here I find more plant species growing in one place than I have seen anywhere else so far. I would not have thought it possible for so many plants to survive alongside one another without some outcompeting the others. I squat down to get a better look, and see, in the small patch around my feet, yellow rattle, bush vetch, ragged robin, self-heal, common eyebright, viola, plantain, red clover, white clover, ragwort, meadow buttercup, lady's bedstraw, and silverweed – and these are just the plants I can identify.

There are other flowering plants, including two species of orchid, and a number of grasses I cannot name. Now this is what I call a wild flower meadow. A pair of tiny blue butterflies dance above the ragged robin; Moss Carder bumblebees are foraging on the self-heal, and the air is full of the distinctive scent of lady's bedstraw. If I could capture just one moment to hold forever as a reminder of our time in the Outer Hebrides, it would be this.

When I finally manage to tear myself away from this little patch of heaven and head back to the beach, Rob is nowhere to be seen. So I walk along the shoreline, collecting bits of seaweed, shells, and driftwood. I find some sea glass, too, and some funny-looking sponges. I love beach combing. I always have, and always will. When I look back at photographs of family holidays to the beach, the ones of me are always head down, searching the shoreline for treasures.

There is still no sight of Rob, so I wander back to sit on a sandy bank at the top of the beach. I choose a spot with a little vegetation, but as I go to sit down, a basking insect flies up from that exact spot, so I move my perch a little farther on. Insects often return to the same spot, so I don't want to be in its way when it comes back.

I look out to sea and am grateful that today the weather is good. There is a heat wave in the United Kingdom, but we have had only a few days without rain since we arrived on the islands, and I can count on one hand the days we have been able to walk or cycle without wearing warm jumpers, scarves, and earmuffs (me, that is, not Rob; he doesn't wear earmuffs). Cold, wet, windy weather is fine for people, but not so good for flying insects. No wonder there are so many of them out and about today. They are making the most of the sunshine.

The insect I disturbed has come back again. I saw it from the corner of my eye as it flew in and landed, but when I look I cannot find it. This makes me curious, so I lean over and scour the area. It must be here somewhere. It is only because I am searching for the missing insect that I notice the holes in the sandy bank. They are small, and there are lots of them, spread out over quite a large area, both above and beneath the bank.

Could this *possibly* be what I think it is? I crouch down on my belly, lying across the rocks below the bank, to get as close as I can. I so want this to be a solitary bee nesting site. No sooner do I make my wish than I see the first bee, leaving her nest and flying straight up and back towards the machair. As she leaves, another lands, the scopal hairs on her legs laden with bright yellow pollen. I can hardly believe my luck. I have stumbled upon a nesting aggregation of Northern Colletes bees (*C. floralis*)! I was hoping to see this solitary mining bee at some stage during our travels. It is found only here in the Western Isles, and in Ireland, where it is widespread.

I spend a happy half an hour watching and photographing the bees flying back and forth to their nests, until Rob returns from his wanderings and we decide to move on. The day has thrown up a number of unexpected treats, and there is one more still to come, for just before we get to the causeway, a beautiful short-eared owl crosses the road right in front of us and perches on a fence post less than twenty metres away. There it sits for a good twenty minutes or so before flying off in the direction of the dunes. I am so glad I agreed to this detour to Baleshore.

As we travel down through Benbecula and South Uist (where we see still more hen harriers and short-eared owls), I become aware that the colours of the machair, and indeed of the entire landscape, are shifting and changing. No longer is yellow the dominant colour. The kidney vetch has gone over, as have the ragwort, yellow rattle, and much of the lady's bedstraw. They are being replaced with pinks, purples, and blues.

By the beginning of the final week of July, the roadside verges and ditches are thick with thistle, the knapweeds are in full flower, and there is clover everywhere. Of all the flowering plants we have seen so far, it is these three – thistle, knapweed, and clover – that seem to attract the highest numbers and greatest diversity of insects. We pull up by a patch of creeping thistle one afternoon, and I count dozens of different bumblebees, butterflies, hoverflies, sawflies, beetles, and other flying insects whose names I do not know. It is a shame that people don't like creeping thistle. It is such a wonderful plant for pollinators.

The day we arrive on the Isle of Barra happens to be our first wedding anniversary and I cannot think of anywhere in the world I would rather be to celebrate. Barra is, they say, the jewel in the crown of the Outer Hebrides, and with its wide-open beaches, mountains, inland lochs, sand dunes, and machair, it really does appear to have it all. We fell in love with this little island on our brief visit three years ago, and we fall in love with it all over again today.

We park up on an old, disused jetty on the Eoligarry peninsula, in the northernmost part of Barra, and use our bicycles to get around for the next few days. But before we set off to explore the rest of the island, Rob suggests we take a walk along the path that leads through the dunes above the beach, and here, just a stone's throw from the jetty, we find Great Yellow bumblebees. Lots of them, foraging on knapweed and red clover. We see White-tailed bumblebees, too, on the white clover, and Garden bumblebees and Moss Carders. In fact, there are bumblebees everywhere.

This, you see, is 'Valley of the Bees' – my name for this special place. It was here, in this very spot in June 2015, that we caught our first glimpse ever of a Great Yellow bumblebee queen – a visitation cut short by the whistle announcing that our ferry home was readying to leave.

The landscape looks very different today to how it looked then. Today it is mostly pinks and purples; the last time we were here, the valley dazzled us with the sunshine yellows of kidney vetch, lady's bedstraw, and bird's-foot trefoil. The bumblebees are different today, too. They are female workers, busying themselves with their pollen-collecting activities. We also see some males, which suggests new queens will soon emerge (if they haven't already).

Next we come across an extremely active nest of Garden bumblebees (*B. hortorum*). It is drizzling, but I sit on the spiky marram by the entrance, mesmerised by the sheer number of workers flying back and forth. I notice a bumblebee hoverfly mimic camped at the entrance of the nest, too. Her colouring is identical to that of the bumblebees. They seem not to mind her wandering in and out of the nest every now and then. I am wet through now, but entirely content. And just when I don't think our day can get any better, I hear the cry of a curlew.

Our adventure is nearly over, and tomorrow we must leave. We sit in our van watching the tide go out, and I know I will hold these islands forever in my heart. Outside the curlew calls again. This is the sound I will miss above all; hauntingly beautiful and other-worldly, it embodies for me the wildness, the very spirit, of these islands. The curlew lands on the beach and I follow it with my eyes, up to the moment a flock of oystercatchers disturb it and it flies off. There are rich pickings here, so surely it will be back before too long.

I will miss the oystercatchers, too, with their bright orange Pinocchio beaks and funny, piping sounds; and the red shanks, ringed plovers, and hen harriers. Strange. I have never thought of myself as a birder, yet it is the birds, rather than the bees, that I have most enjoyed during our time on the Outer Hebrides.

These thoughts are interrupted by Rob, who is making eyes at me from the other side of the van. He is beckoning me over, indicating with his hands that I must move very slowly and very quietly. He nods gently in the direction of the rocks just beneath the jetty, and there, between the sea and the shore, less than ten metres from the van's window, are three snipe, two adults and a juvenile, completely oblivious to our presence.

Snipe are extremely shy, secretive birds, so it is most unusual to see them out in the open like this, although I have to say they are very well camouflaged against the seaweed that they are foraging on. They are quite dumpy-looking, with short legs and long, straight bills, which they are using to probe the sand and seaweed for, I imagine, snails, crustaceans, and insects. I can't believe they haven't noticed us.

Rob slowly passes me the binoculars so I can get a better look. I hold my breath as I lift them to my eyes bit by bit, aware that any sudden movement might alert the snipe to our being here. *Wow.* Now I can *really* see them. The first thing I notice is their beaks, which are long and slender, then their intricately patterned plumage, mostly browns, but with white stripes down their backs. If I take my eyes off them I lose them, as they merge with the rocks and seaweed. I am completely entranced by their understated beauty.

For nearly forty minutes we watch the snipe feeding, preening, resting, and feeding again, before they finally take off and leave us to our dinner, which has gone stone cold. But who cares about dinner when you have just had the birding experience of a lifetime.

As the great, eighteenth-century Gaelic poet Iain Mac Fhearchair wrote in the song 'Smeòrach Chlann Dòmhnaill':

'S I 'n tir sgiamhach tir a'machair,
Tir nan dithean miogach daithe,
An tir laireach aigeach mhartach,
Tir an aigh gu brath nach gaisear

'Tis a beautiful land, the land of the machair,
the land of the smiling coloured flowers,
the land of mares and stallions and kine,
the land of good fortune which shall never be blighted.

CHAPTER 13

On Bovey Heathfield

A s my interest in bees has grown, so has my awareness of everything that surrounds them or connects them to the web of life they exist within. I feel as though I have embarked on a never-ending journey, a journey that spirals continuously outwards, gathering momentum and taking on a life of its own as it sweeps up all the wondrous, wild things that fly, swim, walk, or crawl in its wake.

I am no longer in the driving seat. Rather, I am being driven, or led, by some unknown force far bigger than myself and my desire to 'learn about bees'. If I could draw the route of my journey, I suspect it might look a little like a spider's web, dotted here and there with treasures, whose existence I could hitherto not have imagined, let alone grasped, either in my hands or in my mind's eye.

I find myself giving more attention to some of the wild things that catch my eye than I do to others. Hares, for instance, whenever I see them, take my breath away, as do barn owls. Encounters with both of these animals are, for me, few and far between, which perhaps makes them all the more magical. I am also drawn to the everyday magic of the wild flowers, mosses, and lichens growing in the Dorset lanes around our house, and tantalised by the songs and calls of unknown birds that sing, perched just out of sight, in the trees and hedgerows along these lanes.

I feel compelled, each year as the hedge woundwort flowers, to search out the metallic shieldbugs that live amongst its leaves. I never tire of watching these pretty little bugs as they shape-shift through various larval stages, from flightless green nymphs with smart, black collars and buttons to the splendid adults bearing their very own, copper-coloured coats of arms. And I spend far more time than I should trying to identify and name the moths our moth trap attracts before we release them.

To date, though, I have consciously resisted the temptation to be sucked into the worlds of all the non-bee species whose paths I cross in anything more than a 'Gosh, that's interesting, maybe one day I'll find time to study it in more detail' kind of way. But all that changed last weekend when I spent an afternoon on Bovey Heathfield Nature Reserve, situated on the edge of Dartmoor, with my friend John Walters.

John is a wildlife illustrator; it is he who provided the stunning illustrations in this book. He draws and paints his subjects directly from life, in the field, which is undoubtedly why his portraits of these creatures burst with such vitality. But John is not only an illustrator – he is also an entomologist, and more than that, a speaker, writer, and teacher; long-tailed tit watcher; beetle identifier; and surveyor of the very rare and wonderfully named Horrid Ground-weaver spider (*Nothophantes horridus*). John also happens to know more about the Heath Potter wasp (*Eumenes coarctatus*) than pretty much anyone else on the planet.

Were it not for the fact that I avidly follow John's posts on Twitter, I might never have heard of 'potter wasps', let alone travelled down to the heathland of South Devon in the hope that I might catch a glimpse of one of these incredibly creative creatures.

Bovey Heathfield ranges across fifty-nine acres not far from the small market town of Bovey Tracey. It is the site of an important battle in the English Civil War, the Battle of Bovey Heath, which took place in January 1646, and was designated a Site of Special Scientific Interest (SSSI) in 1989.

But this precious heathland, which once extended across a thousand acres and covered the whole of the Bovey Basin, also happens to be a haven for rare plants and wildlife. According to the Devon Wildlife Trust, it is one of the best remaining examples of the heathland landscape that once dominated this part of the South West. It is most fortunate, then, that what remains of Bovey Heathfield is now owned and managed by the trust.

I had been trying for some time now to arrange a field trip to this heathland with John, but life kept getting in the way. Thankfully, an opportunity has presented itself, albeit at short notice, that coming weekend. We were already hatching a plan to visit my son and his family in Cornwall when Rob suggested we might consider taking a detour to search for John's potter wasps. The weather was due to be warm and sunny – ideal for seeing potter wasp activity – and most important, John just happened to be free on the afternoon of our proposed day of travel. We had arranged to meet him at 2pm, on the edge of the heath.

John's directions to our rendezvous location, when they arrived, confused me a bit. 'Keep going past Mole Valley Farmers and turn left after the zebra crossing by the shops. Drive through the industrial estate to the end of the road and park there,' he had told us. *A nature reserve on the edge of an industrial estate...* That was not quite what I was expecting. Undeterred, we set off that Saturday morning with packed lunch, walking boots, and camera, and arrived at the industrial estate a few hours later.

We drive through the estate to the end of the road, parking as instructed on the junction of Dragoon Close, Cavalier Road, and Fairfax Road, the royalist names that are a nod to the historical significance of the area. I am too distracted by our surroundings to recall my school lessons on the Civil War, however. We are confronted by tall wire fences, shipping containers, parked-up lorries, and pavements littered with discarded fast-food wrappers.

Not in my wildest dreams would I have pictured a site such as this, sandwiched between an industrial estate and a major trunk road, as a haven for rare plants and wildlife. It just goes to show how wrong you can be about appearances, and proves, as if it needed proving, that you should never judge a wildlife site by its neighbours.

We have just finished our picnic lunch and are in the process of donning our walking boots, when John arrives. Having said our hellos, established that the weather is perfect for potter wasp activity, and told John, more than once, how excited I am, we follow him along the side of a large warehouse, towards the gate that will lead us off the industrial estate, and into the wondrous world of Heath Potter wasps.

Lowland heathland habitats such as Bovey Heathfield are becoming increasingly rare and threatened, as are many of their inhabitants. Bovey Heathfield itself is home to no fewer than sixty notable, endangered, or protected species, including, until quite recently, the Narrow-headed ant (*Formica exsecta*). This wood ant is extremely rare. Apart from a number of strongholds in the Scottish Highlands, the only other place it is found in the United Kingdom is at nearby Chudleigh Knighton Heath. John is helping the charity Buglife with its efforts to bring this ant 'back from the brink', including plans to reintroduce it to Bovey.

The heath is also home to adders, grass snakes, and common lizards; ground-nesting birds, such as stonechats, yellowhammers, and linnets; and, occasionally, Dartford warblers. The last are very shy, so you will see one only if you're really lucky. There are other rare insects, too, with wonderful names such as Kugelann's Ground beetle (*Poecilus kugelanni*) and Bog Bush cricket (*Metrioptera brachyptera*).

Alongside the acid-loving grasses like bristle bent and purple moor grass that thrive in habitats like this, lowland heathland is mostly dominated by gorse and heather. The first thing I notice as we walk through the perimeter gate are the colours. What a contrast to the drab, grey lifelessness of the industrial estate immediately behind us.

In the blink of an eye, the landscape has completely transformed, as though we have stepped into a wardrobe and then out through the back into a late summer Narnia. We are surrounded by vibrant yellow gorse, its vanilla-scented blooms sitting atop their prickly, dark green stems. These vie for attention with the equally vibrant, but less prickly, purples of heather and bell heather. Here and there, outcrops of grey stone and bleached, baked earth break through the plants' colours, adding to the impression of patchwork. Gorse and heather, when in full flower like this, make glorious companions. At any other time of the year, the colours might have been muted or scorched, but this is August, and heathland in August truly is a sight for sore eyes.

I am used to being completely dwarfed by gorse, so am surprised that these shrubs appear so stunted. I had naively assumed that all the gorse growing in Britain and Ireland was the same species. Not so. Most of the gorse that grows on Bovey Heathfield, which barely reaches above my knees, is western gorse (*Ulex gallii*). Unlike the mighty European or common gorse (*U. europaeus*), which flowers from January to June, and then sporadically for the rest of the year, western gorse flowers only in the late summer and early autumn.

Rob and I follow John as he winds his way through the undulating heath-land. There are some wide, well-used, parched-earth tracks, but mostly we walk single file through areas of low-growing shrubs. The gorse is very spiky, so I am glad I decided at the last minute to wear jeans. I bet Rob is, too. We are heading, John tells us, for one of a number of exposed areas where, he knows, Heath Potter wasps are likely to visit in order to quarry the clay they use to build their nests. As he has been surveying the potter wasp population on Bovey Heathfield for eight years now, since 2010, there is very little John doesn't know about this species and how it behaves.

As we walk, John talks about some of the other creatures that live and breed on Bovey Heathfield. He points out a Wasp spider (*Argiope bruen-nichi*), which looks like a brightly coloured wasp with eight legs. I crouch down in the gorse, *so* glad to be wearing jeans, and get as close as I possibly can without disturbing the web, to photograph this magnificent creature. As I focus my lens on the spider's eyes, I am struck by the thought that, ten years ago, the very idea that such a thing as a 'Wasp spider' lived on this heath might have put me off coming here. Yet here I am, crouched down in the middle of a patch of prickly gorse, on the edge of an industrial estate in South Devon, marvelling at the splendour of this spider.

What a place. We have been here for barely ten minutes and already I love it. I came here to see potter wasps, but John has filled my head now with the possibility of seeing all manner of rare and exotic-looking beetles, butterflies, ants, and, I cannot believe it, nightjars. *Nightjars live here.* And John knows where they nest. Not that we are likely to see them nesting, for the nesting season is now just about over, but how exciting just to be somewhere they breed.

A part of me is deeply envious of people like John who have been immersed all their lives in the natural world. I wonder how many heathlands I might have walked through in my lifetime without appreciating them as the unique, wildlife-rich habitats they are. Would I even have known they *were* heathlands, rather than moorlands, which are similar but different? Probably not. Yet, I am fine with this, because I have come to relish the upside to not knowing things: There is an enormous amount of fun and enjoyment

to be had when you discover new things. I am (mostly) comfortable in the knowledge that it is never too late to experience connections to the natural world. Once you open your eyes, heart, and mind to your surroundings, you cannot help but notice what has always been there, living under your very nose, just waiting to reveal itself to you.

We arrive presently at the first of John's 'quarries'. It is a small, exposed patch of yellowish clay – flat, dry, and no larger in area than my kitchen sink. I would have walked straight past it. The sun is out and it is warm, so if Heath Potter wasps are using this particular quarry, it will only be a matter of time before one arrives to collect clay.

Whilst we wait, John tells us a little about the recent history of the heath. Until Devon Wildlife Trust bought it and fenced it off fifteen years ago, it had been used for fly-tipping and off-road cars and motorbikes. And already it has breeding nightjars. This just goes to show how successful restoration and conservation projects like this can be.

After ten minutes or so with no potter wasps showing up, John suggests we go and check one of the other quarry sites. The clay at the next site is reddish in colour. When I ask why, John explains that he has created a few artificial quarries, using brought-in clay, so he can more easily observe the wasps. Ten minutes later there is still no sign of a potter wasp, but we see a pair of courting Grayling butterflies (*Hipparchia semele*). This is a first for me – not just the courting, but the Graylings themselves. The courtship sequence is captivating, involving a complex series of short manoeuvres by the male, who alights next to the female and walks around her, raising and lowering his forewings to alternatively reveal, then hide, his dark eyespots. Finally, they face each other, antennae touching. She is impressed. He walks around her again, repeating his moves, and they mate.

We continue our search for about another hour, completing a few circuits of the quarry sites. We are not the only people on the heath this afternoon. There are one or two dog walkers, and we meet a couple who have come to the heath, like us, to look for wildlife. They are friends of John's, so we stop and compare notes. Frustratingly, we learn that they keep seeing potter wasps, whilst we keep missing them.

As it is looking unlikely that we will see a potter wasp quarrying for clay today, John takes us instead to a number of sites he has marked where earlier females built their tiny clay nests for the season. Although I have seen lots of John's photographs and paintings of these clay pot nests, seeing them in person makes me want to clap my hands and squeal out loud. But I contain myself, and restrict my appreciation to making more grown-up remarks, such as 'Wow,' 'Oh, my goodness,' and 'I can't believe a wasp made these!'

The pots are attached to woody stems of heather and each one is unique. All follow the same basic design: perfectly round and jug-shaped, with a small lip and an opening at the top. Some are shorter, squatter, or more textured than others, however. The colours of the pots vary, too, depending on which quarry the wasp has used to collect her materials. They remind me of a beautifully crafted African pot I have at home whose shape I find so pleasing. I am in awe that an insect capable of creating such an exotic-looking nest lives here, in the United Kingdom, on the edge of an industrial estate.

We meander back towards the first quarry, passing, on our way, a pile of dead and rotting-down wood. It is half hidden behind some shrubs and John tells us he often sees nightjars here, perched on the top of a dead tree stump. I can just imagine them, wonderfully camouflaged, their mottled, grey-brown plumage blending in seamlessly against the dead-wood background. You would be hard-pressed not to take them for a part of the stump. Moments later, we see a nightjar ourselves, flying for cover into the trees on the edge of the heath. I do not feel I can claim this as a proper sighting, because all I see is the flapping of wings and the rear end of a largish bird as it disappears into the thicket. But my heart misses a beat anyway.

The sun is at its hottest now, so when we get back to the first quarry, we sit down for a rest. I switch on my camera again, remove the lens cap, and check the settings, just in case a Heath Potter wasp arrives. But it is almost

5pm, and I am quite content with our sightings of Grayling butterflies, Wasp spiders, and the nightjar. I don't need to see a potter wasp today. That would be greedy.

John tells us that Heath Potter wasps continue to be active throughout September and, weather permitting, into early October. We will just have to find an excuse to come back for another visit.

It is whilst Rob and I are discussing this return visit that John calls our attention to the Heath Potter wasp that has just landed, right in front of us, on the clay. I hold my breath. She is extremely striking: jet black with bright yellow markings, and far bigger than I expected. John lends me his binoculars. She flies close to the ground, back and forth across the quarry area, stopping every now and then to examine and test the clay. John says this behaviour is characteristic of an individual that has found a nesting site, and is checking her chosen quarry and water supply before starting to build. She will only use this one quarry, and one water source during her entire two- to three-month lifespan.

She has a very distinctive shape, this wasp. She reminds me of one of those curved, sausage-shaped balloons you get at children's parties, twisted in places, with bulges between the twists. Her abdomen is shaped like a pear drop, swelling out from her long thin waist and tapering to a sharp, pointed, downwards-facing tail. Despite being in the same family (*Eumenes*), she bears little resemblance to our Common wasp, *Vespula vulgaris*.

Just as we think she has begun to quarry the clay, our wasp changes her mind. As suddenly as she arrived, she is gone. We wait a while longer, but she does not return.

———

The following evening, and I have just received an email from John telling me there are now two wasps at the last quarry, and that the one we saw yesterday – our wasp – is building a pot. He managed to track her down late in the afternoon, just before the weather became too cool for her to carry on building.

A few days, and he emails again, this time with photographs of the finished pot and some amazing action shots of our wasp bringing caterpillars to provision her nest after she has laid an egg in it. There are at least three potter wasps active at the same quarry now, and apparently they will carry on building through September. Oh, how I wish I lived closer.

What little I have seen of the Heath Potter wasp has got me, lock, stock, and barrel. I am now every bit as captivated by this species as I am by the snail-shell bees I love so much. I want to know more about Heath Potter wasps: how they 'throw' their pots, their life cycles, and their foraging preferences, including which caterpillars they like to catch to provide for their young. And I have not yet witnessed a potter wasp actually building her nest. I would love to know more.

I email John to ask if he can describe the process to me, and he kindly sends me the account in his nature journal detailing the pot-making process, which begins in earnest after the female has selected her quarry area and water source.

'She flies back out onto the heath, stopping frequently to examine stems of heather, gorse, and dead grasses. Once she has decided where she is going to build her pot, which can be anything from 1 to 120 metres from her quarry, she cleans the stem with her jaws and then wets it with saliva, if necessary, to remove any loose material,' he writes.

So she starts with a clean slate then, or in her case, a clean stem. What next I wonder. 'Having prepared the plant stem, the wasp goes back to her water source to collect a few drops of water before flying on to the quarry, where she starts scraping at the clay with her jaws,' John's account explains. 'Adding water to the dry clay she is able to make a small ball of mud in just a couple of minutes. She transports this ball of mud to the construction site between her jaws and front pair of legs, where the building of her pot commences. This process is repeated, with occasional trips to collect water, between sixteen and twenty-eight times, until the pot is completed. The whole process takes between two and three hours, though this can be spread over several days if weather conditions are poor.'

I am amazed at the diligence the potter wasps display. It is not only 'busy bees' which deserve our admiration, it seems.

John's notes continue: 'After the neck and lip of the pot have been completed, the wasp rests in the nearby heather for a few minutes before returning to lay an egg, which is suspended inside the neck of the pot on a strand of silk. Laying the egg takes about two minutes. After laying her egg, the female potter wasp flies back out on to the heath to search for small caterpillars, preferring those of Pug and Horse Chestnut moths, which feed on heather and gorse. Between eight and thirty-eight caterpillars are brought back to the pot, usually at a rate of about one an hour.' Thirty-eight?! Goodness. These caterpillars must be quite diminutive, for the wasp to squeeze so many into one pot. I must read on.

'Once the pot is full,' writes John, 'the wasp seals it with one or two more balls of clay, then starts to search for a new location to build her next pot. Occasionally (especially in September), she will build a pot adjacent to the first, and sets of up to eight pots in the same location have been recorded. During her two- to three-month lifespan, each female Heath Potter wasp builds about twenty-five individual pots.'

John ends by noting the eggs laid in pots built between May and the end of June emerge as adults before the autumn, whilst those laid in pots built from early July onwards will overwinter and emerge the following year.

I thank John for sharing his portrait of the Heath Potter wasps and their beautiful pots, and tell him how much I look forward to returning to Bovey Heathfield next spring. Hopefully, he will be free to accompany us.

CHAPTER 14

In Praise of Trees

R ob and I are just home after our wet but glorious summer on the islands of the Outer Hebrides. These islands now hold a very special place in my heart. I love them for their wildness; for their wild flowers, wildlife, and wild landscapes; and most of all, for their wild and wonderful weather. But there is one thing they lack, something conspicuous by its absence, and that something is *trees*. There are some trees on the islands, including the odd mini plantation, and I believe there are plans afoot to plant many thousands more. But those trees we saw were so few and far between as to prompt us to remark in the first few weeks, each time we saw one, 'Look, a tree!'

By the end of the summer, however, we had become strangely used to the almost treeless landscape, so much so I didn't realise exactly how much I was missing trees until we happened upon an area on the isle of South Uist called Airidh nam Ban. There, over a period of thirty years, one man, Archie MacDonald, and his family have planted a dense, mixed woodland containing over one hundred thousand trees. The sycamore, pine, oak, and rowan, to name but a few of the trees they have planted, span the length and breadth of the family croft from the water's edge right up the hillside to the highest, rocky outcrops overlooking Loch Eynort. It was extraordinary, to be suddenly surrounded by trees, and to see, touch, hear, and feel their presence again. Yet, the most surprising thing of all was hearing, then seeing, a goldcrest.

Goldcrests are Britain's smallest birds. They are specialist insect feeders, and live in coniferous woodland, very little of which exists anywhere else on the island of South Uist. How this tiny little bird found its way to Mr MacDonald's woodland I cannot for the life of me imagine, but as and when more trees get planted on these islands, I daresay it will expand its range. Who knows, maybe its high-pitched twittering will one day become as common on the islands as the songs of the skylark, lapwing, and curlew.

Walking through this unexpected and isolated woodland made me realise that, though I love the wildness of mountains and moorlands, and the vastness of wide open skies and large expanses of water, it is the tranquillity and stillness of woodlands, with their communities of trees, songbirds, and other inhabitants, which I love the most.

I cannot imagine a world without trees. When I think back on all the places I have lived and visited, all the places I have loved, I always associate them with individual trees. I cannot say for certain what some of them were, especially those from my very early childhood, but if I close my eyes, I can see their forms, colours, and textures as clearly as if I were standing beneath them today. Incredibly, my memories also include trees that I was not consciously aware of at the time.

As I write, I am picturing a huge tree that grew in my grandparents' garden in Suffolk, next to a stagnant pond. Its roots were full of mossy nooks and crannies where fairies lived; it was a magic tree. Then there was the tree outside my other granny's home, a pub – coincidentally named 'The Green Tree' – in the village of Patrick Brompton, in North Yorkshire. It was an ancient sycamore, I think. I looked it up on the internet recently and it is still there (as is the pub), exactly as I remember it.

The farthest back I can go is a tree that grew in the middle of a large lawn outside a house we lived in at Ipplepen, Devon, when I was just four years old. It was truly enormous, its trunk big enough for me to hide behind, and it had an old rubber tyre, hanging from one of its branches by a piece of rope, that I used to swing on.

There are others, which considering I have moved house no fewer than forty times so far in my life, are too numerous to mention. If I were to give it serious thought, I do believe I could map my life in trees.

Whenever I put my arms around a tree – which I do because it is the only way I know of to express my heartfelt gratitude and appreciation for all they do – I feel a powerful exchange of energy. This exchange is difficult to describe in words, but I feel in some way connected to something far greater than my rational mind is capable of understanding. In effect, I feel 'supported', and supported I am, as are we all, every single one of us, by trees, for the debt we owe them is immeasurable. Trees are the lungs of the planet, and without them, we, as a species, could not exist.

Most of us know that trees release oxygen into the atmosphere and absorb carbon dioxide from it. They also absorb other airborne pollutants, such as carbon monoxide, nitrogen dioxide, and sulphur dioxide, by trapping and filtering these compounds through their leaves, stems, and twigs. But how might we *quantify* this in terms we can understand? Given that there are over sixty thousand different species of tree on the planet, it is impossible to exactly measure their individual or collective worth as carbon sinks or producers of oxygen. However, according to the *New York Times*, one acre of mature oak absorbs as much carbon dioxide as is produced by 2.7 cars. An acre also lets out enough oxygen for about eighteen people to breathe over the course of a calendar year. That's a lot of carbon dioxide and a lot of oxygen.

But wait, before you rush out to plant your 'one acre', or fill your garden with random trees, it would be prudent to consider some of the many variables involved before choosing exactly *which* trees to plant, and *where*. It is especially important to choose trees appropriate to your local climate and soil, just as you would with any other plant. If you want to be sure you are planting the right trees, the Woodland Trust offers sound advice on which species to choose and where to plant them in the United Kingdom. Similar charities can be found around the world.

In addition to their vital role as carbon sponges, trees provide much-needed shade, habitat, and protection for numerous plants and all manner of wild creatures, from tiny invertebrates, invisible to the human eye, to the world's largest predators. So, as well as planting trees to help mitigate against climate change, you might like to 'multitask' your trees by including species that are good for local wildlife, in which case it is worth knowing that some are capable of supporting many more species than others.

If it is birds you want to provide for, then you might consider trees like crab apples, hawthorn, or holly, that produce large quantities of fruits and berries. For insect-eating birds, choose those that support large communities of invertebrates, such as oak, ash, willow, and birch.

To address climate change and improve air quality, we urgently need a balance of tree species, which will mean planting more fast-growing and slow-growing trees whilst at the same time protecting the planet's existing mature trees and forests. The older a tree, or woodland, the more biodiversity it supports. This is especially true during a tree's twilight years, when the amount of invertebrates and beneficial fungi living on and around the wood increases exponentially.

And trees benefit us and the planet in other ways, too. They help moderate air and soil temperatures; offer protection from wind damage; improve soil fertility; and stabilise the soil, thereby preventing erosion and mitigating floods.

All these gifts trees bestow on us, and more, but it is not only in practical ways that we benefit from the generosity of these noble giants of the plant world. It has long been known that spending time with trees can have a profound and deeply positive impact on our mental and physical health. The Japanese have a name for this – *Shinrin-yoku* – which means 'taking in the forest atmosphere' or 'forest bathing'.

The practice of forest bathing, which involves spending time under the canopy of a living forest, was introduced in Japan during the early 1980s and has since become a popular way to alleviate stress and promote relaxation. It is now one of the cornerstones of preventive healthcare and healing in Japanese medicine. You may well chuckle at the idea of 'tree hugging' as a form of healthcare, but spending time with trees, or even just being able to see them from your window, has been proved by scientists to help reduce blood pressure, improve sleep, boost the immune system, and accelerate healing from illness. And it goes deeper.

Opening our senses to nature, without any attachment to the outcome, helps us to connect in new ways to the world around us. You cannot rush this process, nor can you force it to happen, for nature does not do what we wish it to do on demand. We have become too used to things happening on a compressed timescale, but the natural world, and trees, especially those that have been around for many hundreds of years, do not work to the timescale of a person's lifetime, let alone a modern person's day.

You cannot sit with your back against an ancient oak for five minutes and expect it to heal you, put your mind at rest, or share the mysteries of the universe with you. But, if you are patient, and if your heart and mind are open, you may in time begin to experience an increased sense of contentment, a deepening friendship, and a feeling of reconnection – not only with your chosen tree, but with the planet itself. I know I have.

When I look back now, I am able to recognise numerous moments when I benefitted from immersing myself in the natural world, though neither the 'benefitting' nor the 'immersing' were conscious intentions at the time. Like others of my generation, I spent my childhood happily playing outside: making dens; playing hide-and-seek; creating miniature gardens in old cake tins; and catching minnows, from nearby brooks and streams, or butterflies, spiders, and other creepy-crawlies from the garden, to keep in my 'zoo'.

My zoo was in one of the outhouses behind the house we lived in on the Malvern Hills sometime in the early 1970s. It was dark and smelled of coal, but the creepy-crawlies seemed to like it. I kept the butterflies, mostly Red Admirals and Peacocks, in the dining room, which we must hardly ever have used to dine in; otherwise, I can't think how I would ever have been allowed to stack huge branches of buddleia on the table for the butterflies to feed on.

Then, because my youngest brother was big enough to have baths in the real bathtub, our mother gave me the moulded plastic baby bath the four of us had all grown out of, so I could collect newts, frogs, and frog spawn from the old quarry lake on the Malvern Hills and put them in it. These are all treasured memories.

My earliest memory of feeling supported by the natural world came between the ages of nine and twelve. My parents were then living in Germany and I was sent to be educated at a convent school in Yorkshire. I was bullied relentlessly. I did have friends, but they were mostly day girls or weekly boarders, so at the weekends, I had no one to hang out with, or protect me. Weekends were, on the whole, pretty horrific.

Yet, come rain or shine, Sunday afternoon always brought a welcome respite. As soon as lunch was over and cleaned up, all of us full boarders would change into clothing appropriate for the outdoors and off we would go for a great long walk – along the banks of the River Swale, up to nearby Richmond Castle, over the moors, or deep into the Yorkshire Dales, depending on the nuns' fancy.

I always walked alone, hanging back from the rest of my class, but I can honestly say I didn't ever *feel* alone. I loved those walks, whatever the weather, and wherever we went. I loved the wind, rain, snow, and sunshine; the river, rocks, trees, tracks, vegetation, and sky. And I didn't merely love them; I also noticed them.

I especially loved the walks where we had to scramble upriver, crossing where it was shallow enough not to wash over the sides of our wellington boots, and climbing over huge rocks and boulders, then up steep banks on the other side, until we finally reached the fields at the top of the valley. I think there were waterfalls, too, but that might be my memory playing tricks. I pretended I was an explorer, and memorised the locations of caves and over-hangs where I might find shelter if I ever plucked up the courage to run away.

I never did run away, though. I can't remember exactly how, but in my early teens things suddenly got better, which meant I began to walk in groups with the other girls. Interestingly, at around the same time as I began to develop new friendships with some of the full boarders, I stopped noticing my surroundings. I wonder, is this how the disconnection process began? I had yearned so desperately to fit in that, when it finally happened, and quite suddenly and miraculously I did fit in, I think perhaps I might have responded by closing the door on nature, a bit like somebody who closes the door on a loyal old friend when someone new and exciting comes along.

I had completely forgotten all this until my marriage of twenty-odd years came to an end about a decade ago. I found myself again spending hours and hours walking, this time aimlessly, on the hills and in the wood-lands. It was a dark time in my life, dominated by feelings of sadness and bewilderment, but as I walked I found strength and peace, and I began to feel calm again. Neither the strength nor the peace lasted long once I came back indoors, but this was the beginning of some new kind of 'knowing' on my part, and a belief that the feelings I experienced whilst out walking might be persuaded to stay with me, and perhaps anchor or balance me, if I could only work out how.

As it turned out, the 'how' was easier than I expected. It gradually dawned upon me that I was walking only to escape life's realities, using the walks as a kind of quick-fix to paper over the cracks. Once I had worked this out, I began to walk every day, regardless of how I was feeling. It was only a little shift, but the difference it made was life-changing and deeply empowering. I began to look forward to my walks and to enjoy them more than anything.

I began to notice sounds, sights, and scents that I had previously been oblivious to. Then I started to notice I was drawn to certain places, and certain trees. On dry days I would take my shoes off and lie down on the side of the hill, just behind a gnarled old silver birch, with my bare feet and hands touching the earth. Sometimes I would sit on the lower branch of the birch, my feet dangling just above the ground, and imagine I was invisible. Then one day, having never hugged a tree before in my entire life, I put my arms around the silver birch and hugged it. And I felt completely, totally, and absolutely hugged back.

The term 'tree-hugger' is used mostly in a derogatory way, or to gently poke fun at people who care passionately about the environment, dress like hippies, wear sandals, and have vegan or vegetarian diets. I'm guessing most of those who use it in a less-than-complimentary way have no idea of its origin. The concept as we think of it today actually originated back in 1730, when 294 men and 69 women belonging to the Bishnoi branch of Hinduism decided to protect the Khejri trees in their village from being used as raw materials to build a palace for the maharaja of Jodhpur. These first tree-huggers were slaughtered as they clung to the trees. However, they did not lose their lives in vain, because their deaths resulted in a royal decree prohibiting the chopping down of trees in any Bishnoi village.

The triumph of the Bishnois inspired the Chipko movement – *chipko* meaning 'to cling' or 'to embrace' – in northern India in the 1970s, when a group of peasant women formed a physical barrier around a group of trees set to be chopped down, by circling and hugging them. The movement quickly spread throughout India as a non-violent protest against the abuses of logging and the destruction of that country's forests, and led, ultimately, to deforestation reforms and a moratorium on tree felling in the Himalayan regions.

I knew none of this when I hugged my first tree, but knowing it now, I would be proud to be thought of as a tree-hugger. That trees bring out such protective instincts in human beings is a testimony to the attachment many of us feel towards them. I do not personally know anyone who would be prepared to die to prevent the felling of a beloved tree, but I know many who work extremely hard, and put their livelihoods on hold, to protect them. Indeed, for the past several years, campaigners have been fighting to save the trees of Sheffield, where 17,500 trees – half of all the street trees in the city – have been condemned to be chopped down by the city council. Most of these trees are healthy and in their prime.

Closer to home, in Shaftesbury, we have our very own dedicated tree group. Fortunately for our local trees, the members of our tree group have not had to deal with threats like those in Sheffield. Instead, they are involved more in raising awareness of the history and whereabouts of important local trees. Having said that, there are trees still standing which might have been felled had the group not been alerted to their plight and saved them. They did so by obtaining legal protection in the form of a 'tree preservation order', or TPO.

To be granted a TPO, a tree needs to fulfil specified criteria, including a level of visibility to the general public; certain age, size, and form benchmarks; having future potential as an amenity; local or historical significance; and rarity of species. None of the TPO criteria surprise me, but I was extremely surprised to discover from the founders of the Shaftesbury Tree Group, Sue Clifford and Angela King, that before a TPO can even be considered, the tree in question needs to be deemed at immediate risk from active felling or damage from development on-site. This means that groups campaigning to save trees often have only a few precious weeks in which to gather the support and evidence required.

Sadly, even if the tree meets all the criteria, it is often still not enough to save it. As has been witnessed through the shocking cull of Sheffield's street trees, the current laws to protect trees are clearly not good enough. Bizarrely, the law affords more protection to listed buildings than it does to living trees.

Sue and Angela remain deeply invested in the Shaftesbury Tree Group, helping to arrange gatherings every couple of months at the Friends Meeting House in the centre of town. The group host talks and events, and organise regular 'tree walks', inviting locals and tourists alike to be introduced to Shaftesbury's 'significant trees'.

I know our local trees quite well by sight, but I love to join in these tree walks so I can listen to Sue, or others, tell us about them – their different common names, and why they are named differently in different places, and other cultural aspects relating to how they came to grow here. The walks usually start outside the Abbey Gardens on Park Walk, which is lined on one side with a row of sycamores. These sycamores were planted in the latter part of the eighteenth century, when it was fashionable to create shaded promenades. Unlike many trees, sycamores cope really well in high winds, which they need to here, given Park Walk's elevated and extremely exposed position above the Blackmore Vale.

Before the walk begins, we stand beneath the sycamores to take in the spectacular views across the vale, and Sue invites me to talk a little about the bees nearby. There are solitary bees nesting in the bank that separates the promenade from the footpath below, and bumblebees foraging on the dandelions that our town council have left to bloom, specifically for the bees.

We leave Park Walk and amble down through the more dappled shade of mature beeches that grow on banks above and below, until we come to Pine Walk. Here, the atmosphere changes as we are flanked by giant, resinous Scots pine. I rarely see bees here, though I have occasionally seen honeybees exploring the pines' deeply fissured trunks searching for resin, which they turn into *propolis*, used to protect their hives and comb from infection and intruders. This is a place for owls and ravens, not bees.

From Pine Walk, we cross the road at the top of St John's Hill and enter Bury Litton. Here, the ambience changes yet again. This is a place of magic and mystery, of secrets yet to be discovered and stories yet to be told. This small, enclosed space is believed to be the site of the old church of St John, all traces of which have long since been obliterated. Here, amongst seventeenth-century headstones, stands the magnificent Shaston Yew, surrounded by ominous-looking laurel (which, until the tree group cut it back, threatened to choke it). Its contorted and twisted branches snake out wide but close to the ground, above the remains of the scattered headstones and hundreds of hidden snowdrop bulbs.

In view of the fact that the national Ancient Yew Group regard it as 'notable', this yew must surely be Shaftesbury's most 'significant tree'. There are more trees to see, but I decide to remain awhile with this glorious tree. As I watch the last of our walking group but me leave the little graveyard, I am struck by the popularity of these walks – a testament to humans' innate love of trees, and the ancient connection we have with them.

European yews (*Taxus baccata*) are normally either male or female, but to know which sex this one is, I would need to see its 'flowers' or its cones, which both sexes produce in the spring, or some winter berries, which would indicate it is female. I have read reports that the United Kingdom's oldest tree, the Fortingall Yew, which has always been recorded as male, has recently started sprouting berries – which basically means it is changing sex.

The so-called flowers on male yew trees droop downwards, and to start with, resemble miniature Brussels sprouts, before developing into cones with insignificant pinkish-white flowers. The female 'flowers' begin life as upright, scaly buds, grow acorn-like with age, then develop into scarlet-red berries that remind me of pimento-stuffed olives, but with the colours inside out. These berries are, by the way, highly poisonous.

Because I have noticed honeybees foraging on the tiny male flowers that yews produce in spring, I rather naively assumed they are pollinated by these bees. In fact, this could not be further from the truth.

Trees are divided into two groups: *gymnosperms* and *angiosperms*. The earliest trees to evolve, first appearing some 390 million years ago, were gymnosperms. This group boasts some of the planet's largest, tallest, and oldest trees, many of them conifers, including pine, spruce, cedar, and yew. The term 'gymnosperm' means 'naked seed' – that is, not enclosed in an ovary. None of these trees produce true flowers, and their seeds, which are directly exposed to the air, are all wind-pollinated.

Angiosperms' seeds are hidden inside fruits. The trees in this group, which all produce flowers, appeared on the planet around 100 million years ago and have evolved alongside the insects, birds, and mammals that visit them for pollen and nectar. Within this group of flowering trees, there are some that still rely predominantly on wind pollination: ash, beech, birch, hazel, oak, and sweet chestnut, for instance. These trees grow mostly in open habitats, where wind pollination is likely to be more effective, and produce elongated (male) catkins that dangle from the branch, ensuring the pollen is easily shaken loose by the wind.

Wind pollination is all a bit hit-and-miss though, so to increase the chances of their pollen reaching a female flower, these trees produce copious amounts of pollen grains. One single cluster of birch catkins, for instance, is capable of producing a phenomenal ten million grains of pollen. This profusion makes them an absolute nightmare for people who suffer from hay fever, but, as you might imagine, extremely attractive to bees and other insect visitors. However, with the exception of willow, which *does* rely on insects for pollination, seeing large numbers of insects collecting pollen from trees with catkins does not necessarily mean these insects are actually pollinating those trees.

Flowering trees that produce highly scented or blowsy flowers, on the other hand – apple, blackthorn, cherry, crab apple, hawthorn, horse chestnut, lime, maple, and sycamore, to name but a few – are typically pollinated, to some extent or another, by insects. Interestingly, not all these trees offer nectar rewards. Some – including lime and hawthorn, in particular – are so fickle and unpredictable that they might produce gallons of nectar one year but almost none the next, as nectar production depends on variables such as soil condition, location, and weather.

Whether or not they rely on insects for pollination, flowering trees are enormously valuable food sources for bees and other insect visitors. On every single foraging trip a bee makes, she uses up precious energy, flying to and from her nest or hive to collect pollen and nectar. If she has access to an abundant source of food, condensed together on a single flowering tree, it stands to reason that she will use up far less energy than she would flying back and forth to individual plants in remotely scattered gardens. The less time a bee needs to stop and refuel to keep herself going, the more time she has to collect food to take back to her brood. So, when you think about it, tree ecology is not rocket science: Planting a flowering tree will help not only tomorrow's climate but today's pollinators as well.

Of all the trees visited by pollinating insects, fruit trees are amongst those most heavily dependent on them for pollination. Apples, for example, rely on insect pollinators, especially native wild bees, more than pretty much any other food crop. Though some apple varieties can self-pollinate, most require cross-pollination with other trees, which means there would be little fruit without insect pollinators.

Nowhere is this domino effect more apparent than in the Sichuan region of south-west China, once one of the largest apple-producing areas on the planet. There, native wild bees have been completely eradicated by a combination of excessive insecticide, herbicide, and fungicide use and loss of natural habitat due to over-extensive farming, and the apples must now be pollinated entirely by human hands.

Every year, thousands of farmers and their families descend on the apple orchards for a few weeks to painstakingly hand-pollinate each and every one of the orchard's billions of blossoms using home-made pollination sticks, some fashioned out of chicken feathers and cigarette filters, which are dipped into plastic bottles filled with pollen. Because apples are a high-value crop, and because labour costs in China are relatively 'cheap', the practice of hand-pollinating is just about viable, for now, but should our pollinators undergo declines like this in other regions of the world, we would be in serious trouble. There are simply not enough humans on the planet to pollinate all our crops by hand.

Trees *are* valuable to the economy, but we must also take heed of their intrinsic value, and the cherished place they hold in our hearts. I feel terribly sad when I hear of trees being referred to as 'natural capital', or 'providing essential ecosystem services'. I abhor these turns of phrase and the way they suggest it is acceptable to determine the value of a tree or woodland in monetary terms. The same wording is often applied to bees when describing their value as pollinators, as well as to the ecological roles of other living creatures. These are not inanimate objects, for goodness' sake; they are extraordinary, beautiful, vibrant, magnificent living organisms, and priceless, in and of their own right.

And it angers me, also, when people talk of 'biodiversity offsetting', as though somehow promising to plant an equal number of trees somewhere else makes it acceptable to destroy old woodlands and the ecosystems that have grown up around and within them. The older a tree, the more it gives to the planet, so whilst planting new trees is good, killing old, healthy ones never is.

Yes, issues like bee decline are being taken more seriously by governments than they might be were pollinators less important to us 'economically', and conservation efforts are often boosted by the income brought in by nature lovers and visitors to national parks and nature reserves, but putting a monetary value on nature itself is, to my mind, a deeply flawed way of thinking. It serves only to increase the growing disconnection between human beings and the natural world.

We cannot have 'economy' without 'ecology'; the two are inextricably linked. And if we don't take care of ecology, that is, our home, Planet Earth, what is the point of the rest? You cannot run a home if your home has been destroyed, no matter how much money you have in the bank.

CHAPTER 15

Sedgehill, a Natural History

The image on my laptop has me well and truly stumped. There was no doubt in my mind, when I clicked the shutter button on my camera, that this was a female solitary bee of some kind, but the creature I am now looking at appears to possess two black claws or pincers – similar to those of a crab, or perhaps the mouthparts of a large beetle. Its body is mostly hidden, buried deep inside the flower, and only the pincers (if pincers they are) are visible, poking straight up towards the sky as though waiting to catch some passing insect.

I don't get it. I know my identification skills leave a lot to be desired, but she was most surely a solitary bee of some kind.

I first noticed her earlier this morning in Diana's garden at Sedgehill where Rob works. I go there sometimes to give him a hand with the weeding, but spend just as much time walking around photographing flowers and bees as I do pulling up bindweed.

Rob has been working this garden for the past fifteen years and it is a testimony to his love of nature that he has managed to maintain a balance between the formal lawns and herbaceous borders, packed with traditional cottage garden shrubs and perennials, and the semi-wild areas, including the orchard and meadow, without once using insecticides, herbicides, or fungicides. Nor has he ever used lawn fertiliser.

As a result, the garden, all two acres of it, has achieved a natural balance and become a haven for invertebrates, slow worms, amphibians, and garden birds. I have counted more different bee species in Diana's garden than I have in any other single space.

The bee I saw this morning was shiny black, and quite striking against the clump of pretty yellow flowers which she kept coming back to. I stalked her for a while through the lens of my camera before snapping her image when she landed, conveniently, in the bloom closest to where I was standing.

So, unless there was something else hiding inside that particular bloom, this must be the same bee. I crop the image further and rotate it. The black 'claws', I can now see, are covered in copious amounts of bright yellow pollen. A pollen-collecting beetle perhaps? I look at it from a different angle, and this time I see membranous wings, folded neatly across one another, the rear end of an abdomen, and the tips of two long antennae. It *is* a bee, then. So what's with the pincers?

Suddenly, it comes to me. I know exactly what species this is. That is, I think I know, but surely I am mistaken. For starters, the habitat in this garden seems all wrong. It is not wet enough. Although there are a couple of large ponds, these are both man-made, and there are no marshy areas that I know of, apart from the area down by the stables that becomes a bit boggy in the winter.

Mainly, though, I am not quite sure of the name of the flower she was collecting pollen from – which, in this bee's case, is going to be fundamental to confirming her identity. I need to ask Rob. If this flower is what I *believe* it is, then the 'pincers' are not pincers, but legs – claw-shaped in appearance only because of how enlarged they are compared with other bees' legs – and the bee I saw this morning is none other than a Yellow Loosestrife bee (*Macropis europaea*).

From what I know about this species, I believe that in Britain it collects pollen only from yellow loosestrife flowers, hence its vernacular name. But the Yellow Loosestrife bee is also unique amongst British species in that the females collect floral oils, some of which they add to pollen to feed to their larvae. More unusual, and most ingenious, is the way they also use these oils to waterproof and protect their underground nests, which are often built in areas liable to flooding such as marshes, fenlands, and riversides. Although there are other bee genera, as well as at least a dozen other *Macropis* species outside Britain that collect floral oils, the Yellow Loosestrife bee is the only British bee that does this.

As with the pollen, the oils are collected in Britain and Ireland by these bees only from yellow loosestrife flowers. To help them soak up the precious oils, which the flowers secrete from glandular hairs situated at the base of their stamens (the male reproductive parts of a flower, which produce pollen), loosestrife bees have evolved specially adapted, silver-coloured hairs on their enlarged hind legs.

However, yellow loosestrife is not a one-stop shop for this bee, for the flowers of this plant do not contain nectaries, and therefore, these bees need to visit other flowering plants to refuel themselves with nectar. Although they are extremely choosy when it comes to pollen and oil, they are far less fussy about their sources of nectar, and can be seen visiting many different flowers, including bramble, willowherbs, mints, creeping thistle, and knapweed, most of which are available in and around Diana's garden.

There is, however, yet another unusual feature about this bee: The females are known, on occasion, to extend their two back legs up vertically in the air whilst visiting flowers. No other bee species in Britain or Ireland does this.

I have spent hours trawling the internet trying to discover exactly why Yellow Loosestrife bees exhibit this behaviour, to no avail. Earlier I came across something that suggested the leg posturing might be the female's way of indicating to passing males that she is 'not interested', but I cannot for the life of me remember where. I did not see any passing males in the garden today, though, so I have no way of knowing if the idea has any merit. It did occur to me that the female might be holding her legs in the air to prevent pollen being lost whilst she collects oils, but I have found photographs of her doing the leg display without her carrying a pollen load. Either way, the fact that my bee displayed this behaviour helps me to confirm her identification.

I open the last of the papers I have bookmarked. 'Foraging, Grooming and Mate-seeking Behaviors of *Macropis nuda* (Hymenoptera, Melittidae) and Use of *Lysimachia ciliata* (Primulaceae) Oils in Larval Provisions and Cell Linings' is the mouthful of a title. It is dated 1983 and its authors are James H. Cane, George C. Eickwort, F. Robert Wesley, and Joan Spielholz. I scan through the summary of the researchers' findings. There it is: 'hind leg posturing to advertise mating non-receptivity are also discussed.' Now you're talking!

I am over the moon to discover that female *Macropis* bees do indeed raise their hind legs in the air to signal to passing males that they are not interested in mating. This is the evidence I have been missing. Still, the paper doesn't explain why *my* female was raising her legs in the air. There were definitely no 'passing males' when I photographed her, but as the authors have documented this behaviour in the field, I am nevertheless convinced. I would love to know more. If only I could speak with one of those authors...

I do a bit of a search on the internet and manage to find the first author listed on the paper. Jim Cane, as he is known, currently works as a research entomologist for the US Department of Agriculture. I have a look at his biography and a few key phrases jump out, namely that he has a 'long-term interest in conservation' and has been 'studying solitary bees for thirty years'. There are a million and one things I would like to ask him.

We email back and forth for a few weeks and finally manage to arrange a telephone call. I already know from our email exchanges that I am going to be fascinated with what he has to tell me – and so I am; he is a mine of information about all things bee.

Jim begins by explaining that the paper I found was a small part of his graduate dissertation, 'all about Dufour's gland secretion, its chemistry across a broad survey of the bees, and its uses by nesting bees', back in the early 1980s. Although he is now based in Utah, it was in New York State that he observed the *Macropis* bees for his research. I ask if he remembers them, and he can.

He says he watched the males 'cruising' loosestrife for females, constantly pouncing upon them (as male solitary bees are wont to do), before being kicked off abruptly by those females that had already mated. The already-mated females 'rejected the pouncing males with a single, synchronous, upward thrust of their extended hind leg', then went back to their business of foraging for nectar, pollen, or oils whilst keeping their legs extended for a couple of minutes afterwards. This could explain why my female's legs were extended even though I didn't see any males at the time.

We chat a while longer. I tell Jim about the *Macropis* species I saw in Diana's garden, and he tells me more about the nesting habits of the *Macropis* species he observed as a student. Their nests are always quite shallow, he says, comprising a main tunnel from which the female digs lateral tunnels, at the ends of which she constructs one or two nest cells. From what Jim and his colleagues observed, the female lays no more than ten eggs in one nest, and to his knowledge does not go on to create other nests.

I wonder how he worked this out, and he explains that seeing worn wings on a female initiating a nest would have suggested she has already constructed a nest elsewhere, but every single one of the bees he watched initiating nests had pristine wings. I also learn that *Macropis* bees usually forage within thirty metres of their nest; do not nest in large aggregations; and that the *Lysimachia* oils the females collect have a greenish tinge that can be easily seen on the walls of the cells after it has been used to waterproof them.

Towards the end of the call, we touch briefly on Jim's passion for conservation. I have been asking numerous question about *Macropis* nesting and mating behaviour, but it becomes abundantly clear that I could find answers to some of these questions myself, simply by planting more *Lysimachia* in the area where I saw my Yellow Loosestrife bee, and spending time observing the species myself.

Jim encourages me to do just that. He stresses the great value of *citizen science*, observations made by non-scientists, in building up a picture of how solitary bees, or any other wildlife, for that matter, behave. Our conversation reinforces my determination to take more notes when I am watching bees, just in case I notice something that hasn't been seen before.

Even more important than recording bee behaviour is recording the fact that these species are *there* in the first place. After all, if we do not know something is there, we cannot possibly know when it has gone. This is something Richard Comont, science manager for the Bumblebee Conservation Trust, is passionate about. 'Biological recording', Richard says, 'especially long-term monitoring, is the very basis of ecology – the study of species and their environments.'

And it is no use keeping this information to ourselves – we need to share it, so that conservationists and scientists can act upon it. It is because of the Butterfly Conservation's world-leading monitoring and recording scheme, which has been running since 1976, that we are aware of the serious, long-term, and ongoing decline of UK butterflies, and able to look at what might be causing this. We can and should all get involved in more citizen science schemes.

Many of us already take part in the RSPB's Big Garden Birdwatch each January, or the Botanical Society of Britain & Ireland's New Year Plant Hunt, but we can make an even bigger difference by sending in records of the creatures, particularly the insects, we see throughout the year. Further, as Richard so rightly points out, 'without reliable species-level IDs, all else is suspect.' Fortunately, it has never been easier to get help in identifying your finds. All you need to do is take a photograph and upload it to the iSpot website, where someone will help you identify whatever plant or creature you have found, and explain where and how to submit a record.

I come away from my call with Jim thinking I can safely assume the leg display behaviour exhibited by both *Macropis nuda* (Jim's bee) and *M. europaea* (my bee) has evolved for the same reasons. But even with all the evidence I have gathered, there is one last, loose end I would like to tie up. And that is the fact that the plant in my photograph does not look like the plant I know as yellow loosestrife.

I send Rob a quick text: 'What's the name of that plant in the top border with the pretty, little yellow flowers,' I ask, 'the one in front of the tree lupin and to the right of the red-hot pokers?' Then, because I am far too impatient to wait for Rob's response, I key in 'yellow loosestrife' and do another internet search.

I immediately recognise the flowers offered up by my search engine as the rampant yellow spires that grow at the back of our own little garden, just behind our pond. '*Lysimachia punctata*', it says underneath the images. But this is definitely not the plant my bee in Diana's garden was collecting her pollen from. In fact, apart from having similarly shaped yellow flowers, it is altogether quite different. I wish I knew plants by their scientific names as well as their common names. It would make my search easier if I could narrow it down by taxonomic family. Hopefully, the plant I'm looking for is in the same genus.

I scroll down a little farther, and bingo – there is another yellow loose-strife plant, *L. vulgaris* – and this variety is identical to the plant in my photo-graph. Compared with the cultivated yellow loosestrife, *L. punctata*, which stands erect like an army of tall yellow spires heavily crowded with masses of clear, yellow flowers – this wild variety, *L. vulgaris*, looks softer, branched rather than erect, and more leafy. It has thin, flowering stalks, and its flow-ers, which are slightly more closed than those of *L. punctata*, are arranged more loosely and in pyramid-shaped clusters. The pale green sepals, where they grasp and enclose the base of the petals, are prettily edged in reddish orange. This plant is more delicate and pretty than the cultivated variety.

It's official, finally. I have seen, photographed, and identified *Macropis europaea*, the Yellow Loosestrife bee. I cannot wait to submit a record of my find to Bees, Wasps and Ants Recording Society (BWARS) and announce it on Twitter. But before I do, I am curious to discover what this bee is doing in

Sedgehill. Call me an anorak, but as I have been so diligent in making sure I have the correct identification for this bee, I would also like to have a feeling for its nesting habits – that is, why it is *here* – before I submit my record.

Things have changed a lot since I first started trying to identify bees. Mercifully, I no longer need to trawl through the BWARS website searching for lookalikes. Not that my identification skills have improved that much; despite my greatest efforts, and considering how much time and energy I put into improving and honing them, they are still woefully inadequate. What *has* changed is that I now have an excellent book to help me out when I am stuck. Steven Falk's *Field Guide to the Bees of Great Britain and Ireland* is my bible. I trust it will tell me more about the habitat and nesting require-ments for these bees, and also something about their status and geographi-cal ranges. If they have not been recorded in this area before, it does not necessarily mean they are not here, but if, for instance, they have only ever been recorded in the farthest northern parts of Britain and Ireland, that would suggest a one-off sighting, an anomaly.

I turn to page 264 of the *Field Guide* and read that the Yellow Loosestrife bee is 'mostly recorded in southeast England from Dorset to Norfolk'. Sedgehill is on the border between North Dorset and Wiltshire, so that fits well. I look to the section on habitat and nesting requirements: 'wetlands and watersides with plentiful yellow loosestrife, including fens, roadbeds, streams, ditches, and canals. Occasionally recorded in gardens.' As I thought, this bee has a preference for wetland habitats, but I am not aware of there being 'plentiful yellow loosestrife' in this area – and there are certainly no fens or canals. Then again, this village is surely not called Sedgehill without some reason.

I am now ready to submit my record to BWARS, but I am also more intrigued than ever about the origins of the name and natural history of this parish and its surrounding landscape. My understanding is that sedges are mostly associated with wetlands, and the fact that *L. vulgaris*, a plant that favours wet places, is growing happily in Diana's garden indicates this area must be wetter than I previously imagined.

I get to work, eventually landing on the 'community history' page on the Wiltshire County Council website. The information I glean there is gratifying: 'The former Sedgehill parish…lies on Kimmeridge Clay, a soil particularly suitable for use as pasture and meadowland,' it begins.

'Variations on the name, such as "Seghull[e]" and "Segghull" were recorded from 1241 and it is likely that the parish name signifies "a hill where sedge grows"' I learn that 'small pools lie at scattered points in the parish whilst three large pools are fed by the Sem.' I do not recall seeing these, then read that 'some of the pools have now been drained.' It seems that 'the south-western boundary with Dorset also follows the course of streams.' There are so many references to water, my head is spinning.

But what started out as a quest to establish how Sedgehill provides habitat suitable for Yellow Loosestrife bees has taken me deep into the history of this area. Beyond finding these references to streams and other watery habitats, I have noticed that the names of local farms have often, historically, referenced water. Most are now long gone, or have merged, but, significantly, I have seen, on a reproduction of a beautiful map dating to 1773, that there used to be a Burybrook Farm, a Whitemarsh Farm, and a Westmarsh Farm in the vicinity. It just so happens that Burybrook Farm – which is still here, farmed by David, and now spelled Berrybrook – and what was once Whitemarsh Farm (now part of Berrybrook) are not too far from the garden where Rob works. And what was once Westmarsh Farm lies just a stone's throw to the north.

This history makes me see Diana's garden in a different light. I realise I was focusing more on the flowers that grow in the borders than on the condition of the soil and surrounding land – and have completely missed and overlooked many (now obvious) indicators of a damp habitat.

It occurs to me that if I were to go looking in the open sunny areas surrounding Diana's garden, that this plant favours, I might find more Yellow Loosestrife bees. I know from the Wiltshire County Council community history page that there is still a fair amount of common land around Sedgehill. I don't need permission from anyone to walk through these pockets of land, or along the local public footpaths and bridleways, so these are where I will start.

Unfortunately, we are heading off on holiday this coming weekend, and by the time we get back in early September, the bee's flight season will be over. I will have to ring-fence the months of June and July next year to tramp the fields, footpaths, lanes, dykes, and ditches of Sedgehill to find these bees and the plants they depend upon.

This experience makes me realise, too, that I have turned a corner in the way I respond to coming across some new insect or plant. Previously, I would have checked the bee's physical features, confirmed its identification on Twitter, submitted a record, and then filed it on my desktop in the 'bee pics' folder. This time I found myself asking questions: *What is it doing here? Where has it come from? Why does it stick its legs in the air like that? What does it all mean?* I have become a nature detective, investigating, digging deeper, leaving not a stone unturned in my search for evidence that the village of Sedgehill might be home to more than just one, lone Yellow Loosestrife bee.

I have enjoyed unearthing all my evidence, and using it to find answers to some of my questions and help make new connections. It has been interesting, too, to learn about the history of Sedgehill. I did not know that it has always been of a dispersed nature, a non-nucleated village which, unlike those that cluster around a focal point like a church or a village green, is made up of smaller hamlets and farms, some as far as two and a half miles apart, and linked by a network of lanes and tracks.

I often wander these lanes on days when I drop Rob off at work. It takes me a full forty minutes to walk once around the village, following a kind of a loop. In the first part of the nineteenth century, there were no fewer than nineteen farms in this parish, but when Rob started working here in 2003, only nine of these farms were left, four of them owned by the council. Now, only three of the five privately owned farms still remain, and the last of the council farms, Berrybrook Farm, is in the process of being sold off.

David's family has been farming Berrybrook since the 1950s. After the Second World War, council policy was to buy up farmland and divide it into smaller farms and holdings that could be let out to young families in a bid to attract a new generation into farming. Council priorities have changed, and they are now under pressure to provide more housing and find funds to replenish services affected by budget cuts. Council-owned farms are just one of the casualties of our unsettling times.

Pockets of land in and around Sedgehill, which for centuries was made up mostly of pasture and meadow, are already replaced with arable land. What will happen once the remaining farmland is sold off? What will be the impact over the next few decades on the landscape and the flora and fauna it supports? If farm buildings are replaced with residential buildings, some of the lingering marshy land in Sedgehill might be drained and paved over, resulting in reduced habitat for the Yellow Loosestrife bees. And council-owned farmland is being sold off all over the United Kingdom. Where all this will lead, and how it will affect the habitat and biodiversity of Sedgehill and other parts of our countryside, only time will tell.

I am becoming increasingly aware of the parallels between my interest in history and my love of the natural world. It is obvious that the two are inextricably linked, not only because of the ways humans have interacted with and changed the landscape over the ages, but also in the similar ways you search for answers to historical and natural mysteries by following a series of clues to build up evidence. It occurs to me that I might have missed my vocation; had I known such a subject as 'natural history' existed when I was younger, I might well have chosen to stay on at school and study it.

———

Studying history is not just about learning the names of kings and queens and remembering the dates of battles. I was in my very early teens when I first realised this. My mother gave me her copy of Josephine Tey's *The Daughter of Time* to read. It was a detective story, about a modern police officer setting out to prove that Richard III could not possibly have killed the Princes in the Tower. The detective had a 'hunch' that the man whose face looked out at him from a well-known portrait of Richard III, the Crookback, was not the face of a murderer, so he conducted an investigation in much the same way as one would with a modern murder, finding and examining evidence, putting together facts from clues, discarding rumours, and eventually coming to a more logical conclusion that it might have been Henry VII who ordered the murders.

I consumed the novel from cover to cover in just one day. Prior to reading it, I had completely bought in to the narrative that Richard III was a cruel and wicked king who had murdered his nephews to gain power. I believed this story simply because it is what I had been taught at school. Josephine Tey taught me how important it is to dig deeper, and to ask questions.

I have not thought about this book, or about 'history', for years, but learning about the history of Sedgehill, and thinking about its future as a habitat, has reminded me of the importance of challenging the status quo and asking questions. I cannot wait to learn more about which plants and insects are indicators of certain habitats and landscapes, and visa versa.

And to think it was a bee that opened these new windows of discovery for me.

CHAPTER 16

Cotton Weavers

It is far too hot to continue working on the allotment today. The parched ground, more like concrete than soil, resists my half-hearted attempts to rid it of the bindweed that has taken hold during the weeks we have been away. Far from pulling this tenacious plant out by its roots, all I can manage today is to snap it off at the surface. I give up.

We acquired this tangled jungle of bramble, dock, and bindweed – an extension to our existing plot – earlier this summer, but knowing we would be away for most of the following month, we decided to cover it with a piece of old carpet until we find the time to tackle it. Apart, that is, from the small area I am supposed to be weeding now. Although this patch was just as overgrown as the rest when we inherited it, we discovered a healthy-looking rosemary growing here that had somehow managed to hold its own amongst the dominant overgrowth which had beaten everything else into submission. It seemed a shame not to try and save it. That was back in June. It doesn't look very healthy now, nor is it holding its own any longer.

Maybe I can still free it from the encroaching bindweed, which has broken cover some fifteen centimetres from the base of the rosemary, crept across the open ground, and then jumped the gap between earth and plant, like bindweed does. It is coming at the rosemary from more than one direction, threatening to completely surround and smother it. If I don't break the bindweed off at the base now, it will soon spiral its way along the lower parts of woody growth before binding with more of its kind to form a thick, strong rope that will eventually strangle and choke its poor victim. Snapping the bindweed off at the surface like this won't save the rosemary in the long term, but I am just going for damage limitation today. It will keep the bindweed at bay until the ground is more workable and I can wheedle out the roots.

You've got to admire bindweed – that is, *you* don't, but I can't help myself from doing so. We have two kinds on our allotment, hedge bindweed and field bindweed. Bees love them both. Hedge bindweed has large, heart-shaped leaves and huge, white trumpet flowers. It climbs and smothers the tall plants and shrubs in garden borders and hedgerows. Field bindweed is much smaller, with pretty, pale pink-and-white trumpet blooms. It seems to spread more than climb, though it still strangles any low-growing plants it can get hold of (if I don't notice it in time). No matter how sure I am that I have eradicated all the field bindweed on our allotment, it just keeps popping up, coming back every single time we clear an area for planting. It is indestructible, irrepressible, and indefatigable. And I would rather not wage war on it.

I often wonder, if we were to let our allotment go, what it would look like five years from now. Would any of the flowers and shrubs we have planted and nurtured still be here? I imagine the buddleia would survive, and the woundwort seems to hold its own without our paying it much attention. But judging by the plot next door, which has been let go for a few years, most of our allotment would revert to bramble, bindweed, dock, and nettle. This says something about the resilience and tenacity of our native plants, as well as the fragility and dependence of those more tender plants that need our support to survive.

I have been thinking a great deal about such things since reading Isabella Tree's outstanding book *Wilding*, where she tells of the nightingales, turtle doves, and Purple Emperor butterflies living and breeding on the farmland she and her husband, Charlie Burrell, have allowed to go 'wild' in Sussex. That these birds and butterflies have appeared at all is a triumph, but it is their numbers, compared with populations in other areas of England that are specifically managed by conservationists for their benefit, that make their presence at Knepp Castle Estate so exciting. Some of these species were previously believed to prefer 'woodland' habitats, but, at Knepp, having been given freedom to establish themselves wherever *they* choose, they have chosen areas of extended hedgerow and scrub. 'Categorising species as "woodland", "farmland", "heathland", et cetera, is a relatively new phenomenon,' says Isabella. 'Look back in the literature of the great Victorian field naturalists and you often find a very different, more nuanced picture. We've forgotten that we live in such a depleted landscape that many creatures are clinging on by their fingernails in habitat that is not optimal – they'd much rather be somewhere else. But we assume that's where they'd "like" or are "meant" to be.'

If Isabella and Charlie had not had the courage to sit back and give nature the freedom to express itself, we might never have known of these, and other creatures' *natural* preferences. Rob and I cannot hope to attract nightingales, turtle doves, or Purple Emperors to our little plot, but my mind has been buzzing since I read *Wilding*, and I am inspired to think about how we might provide more space to give wildlife the freedom to express itself elsewhere in Shaftesbury. For now, though, our focus is on growing food for our table and plants for local pollinators.

I look around to see which other plants are coping with this summer's drought. As our summers get hotter and drier, it is important to note what thrives and what dies during periods like this. Whatever we plant here needs to be robust and fend for itself once it has established. We can neither afford, nor do we want, to become slaves to watering, so those plants that have not coped this year will be replaced next year with more of those that have. The lavenders seem happy enough, as do the salvias, catmints, toadflax, California poppies, and, of course, the calendula. Calendula is always happy. All these flowers are in full bloom and attracting an interesting and diverse variety of insect visitors. The spinach and chard are not doing so well; they have bolted, as have the coriander and all the salads; and most of the wild flower annuals have gone to seed. Down at the bottom of our plot, I am relieved to see how well the lamb's ears (*Stachys byzantina*) is faring. Then I remember this is one of the plants being grown in the Mediterranean Biome at the Eden Project, in Cornwall, so I need not have been concerned for it. Either way, I am both relieved and delighted, because this is one of the plants I particularly want to do well here.

We planted the lamb's ears two autumns ago in the hope that it might attract one of my favourite solitary bee species, the Wool Carder bee (*Anthidium manicatum*). The leaves cover quite a large area, just below the lavender, and have already spilled out across the path, blurring the edges between our allotment and David's. He won't mind. We are most fortunate in our neighbouring plot holders, David and Pilar; not only are they tolerant of our sprawling (and rather unkempt-looking) plot, but they also use very few, if any, pesticides.

Friends farther down the site who, like us, garden with wildlife in mind, are less fortunate. Many of their neighbours are still stuck in the mindset that says you cannot keep a tidy plot or produce a good crop without using liberal amounts of herbicides, insecticides, and fungicides.

My father was of this mindset, and though I tried my hardest to persuade him otherwise, he remained completely closed to the idea of growing without a shed full of pesticides, lawn improvers, fertilisers, and Lord knows what else, till the day he died. He loved his garden and his vegetable plot every bit as much as we love ours, and prided himself, as does Rob, on the produce he grew for the table (especially when he turned that produce into his signature courgette soup or green tomato chutney). My father just had different priorities. And some days, whilst digging up the field bindweed, for instance, I can sympathise with the desire to take a less physically punishing approach.

I always try to remember this approach when I am talking to fellow gardeners. Though I'd like everyone to stop using pesticides today, you cannot force other people to think the same way as you. It's often best to talk about the fact that there are alternatives and then encourage people to reduce or end their dependence on pesticides, pointing out their negative effects on our pollinators and other 'non-pest' species.

Despite the drought and the bindweed, the lamb's ears really is looking fabulous. Last year it produced only a few flowers, but it is more established now. There are dozens of thick, chunky spikes, each standing proud above a carpet of the silvery green, downy foliage that makes this plant so attractive to gardeners. Whorls of tiny pink blooms cluster around the stalks, providing forage for Buff-tailed, Garden, and Common Carder bumblebees; honeybees; and a number of different species of hoverflies. On some of the leaves there are shieldbug nymphs, too. They are the same species I sometimes see on the hedge woundwort that grows up by the cottages, and occasionally on white dead-nettle. I wonder for a moment what the nymphs are doing on the lamb's ears, then recall that hedge woundwort is also a *Stachys*, and that all three plants I have seen these bugs on – lamb's ears, hedge woundwort, and white dead-nettle – are members of the mint family, Lamiaceae. My rudimentary knowledge of taxonomy is coming in handy.

I scan our plot to see if I can identify other plants from the mint family. There are loads. We have betony, bugle, hyssop, motherwort, self-heal, and wild marjoram, plus a number of different thymes, salvias, and catmints, each and every one a magnet for bees. I do so love making connections like these. The child in me wants to share my realisation with someone else, but there is no one around. I smile quietly to myself and carry on admiring the lamb's ears.

I am surprised not only by the numbers but also the huge diversity of bees and other insects visiting this plant. In truth, I had been a little 'species specific' in choosing it for our plot, but I feel wholly vindicated now that I see how popular it has turned out to be with so many other creatures.

The bees I had planted the lamb's ears for, Wool Carder bees, are distinctive in appearance. They are mostly black, more shiny than hairy, with creamy white hairs on the lower segments, or *tarsi*, of their legs. But their most distinguishing feature is the bright buttercup-yellow markings along the sides of their abdomen, some like broken stripes, others more like dots. There are twelve yellow marks in all, six on each side. and they look to me as though they might have been painted on as some kind of tribal war paint, or as an afterthought. Both male and female Wool Carders are quite large compared with other solitary bee species, and this is the only bee in Britain and Ireland whose male is larger than the female.

Ironically, amongst all the insects enjoying our lamb's ears today, which must number upwards of thirty individuals, I cannot see one single Wool Carder bee. So much for the adage 'If you build it' – or in this case, grow it – 'they will come.' We didn't see any Wool Carders last year either, but that didn't surprise me, as lamb's ears takes some time to establish itself. However, looking back, there were woefully few Wool Carder bees anywhere in Shaftesbury last year, not even in the garden of my lovely friend Juliet, whose beautiful, mature herbaceous borders are an absolute haven for bees and other flying insects.

Juliet is a seasoned gardener, with years of experience and a great deal of passion for the plant varieties she selects for her garden. A few years ago, she invited me to run some afternoon bee tutorials and walks for her and her friends, and having learned more about the diverse and fascinating creatures that pollinate her patch, she has herself become a dedicated bee watcher and advocate. She emails me now with news about everything with six legs and two pairs of wings that visits her garden – sightings of Vestal Cuckoo bumblebees, Ruby-tailed wasps, Early bumblebee males, and much more.

Together with the messages I receive from my friends Jane and Jo (one describes butterflies and barn owls in a way that makes me believe all over again in nature spirits and fairies; the other speaks of the insects and birds that visit her mini meadow in Oxfordshire, as well as the water voles and hares that inhabit the wild areas near her home), Juliet's are the virtual scribblings I enjoy receiving most in the world. Having friends who wax lyrical about the joy and magic they experience in their encounters with nature is a precious gift, one I hope I never take for granted.

Juliet and I have been keeping vigil for the Wool Carder bees this summer – she in her garden, I on our allotment. I first told her about this species, and promised they would not be able to resist the extensive areas of lamb's ears growing in her borders, some three summers ago, and we have both been disappointed year after year. I have probably been even more disappointed for Juliet than I have for myself. I had wanted this year to be 'the year'. There is still time for the Wool Carders to appear, though, so I have not given up. Not yet.

A few days later, I receive an email from Juliet. 'Finally, they have arrived!' she writes. I'd better get down to the allotment, *pronto*.

It is warm again, and there are even more insect visitors on the lamb's ears this afternoon than there were yesterday, but not yet any Wool Carder bees. Various bumblebees amble from bloom to bloom without a care in the world. If I could communicate with these gentle bee folk, I would advise them to make the most of the afternoon's peace and quiet, for any day now, this and other patches of lamb's ears the length and breadth of the country will become battlegrounds. And there may be casualties.

In areas where populations of Wool Carder bees make their nests, patches of lamb's ears and other plants, such as woundwort, great mullein, and nepeta are jealously patrolled and defended by large males looking to mate. The females visit these plants to collect plant fibres for their nests, or to forage. The males can be quite domineering and aggressive. This behaviour is called *resource defence polygyny*, and it is extremely unusual amongst bee species of Britain and Ireland.

Armed with lethal weapons in the form of five sharp, mace-like spines that protrude from his tail – three from the end and two from the sides – the male Wool Carder is ready to fight off any intruder, no matter how large, who dares enter his territory. Any flying insect that wishes to take advantage of what this plant has to offer must be prepared to run the gauntlet of this fearsome gladiator bee. Only his harem of female Wool Carders are welcome here, and in return for his services protecting the patch for their exclusive use, he will mate with them.

Unlike most other solitary bee species, female Wool Carders are also polyandrous, mating frequently, so a dominant male is rewarded for his efforts many times over. Smaller Wool Carder males are forced to sneak into a dominant male's territory whilst his back is turned, or he is sunning himself on a leaf, for the chance to mate with a female. If caught, the smaller male will be chased off, or worse.

The patrolling behaviour of large, dominant Wool Carder males is quite fascinating. They dart back and forth, up and down, and in between the stalks of lamb's ears (or whatever plant they have taken ownership of), hovering as they survey their territory, before zipping off to check other plants in their patch. Woe betide any unsuspecting insect that happens to be caught in their territory, for they will hurl themselves, full speed ahead, at the interloper, colliding in mid-air, or knocking it off a flower, and wrestling it to the ground. If the victim is fortunate, it will escape. If not, the male Wool Carders are more than capable of crushing and killing other bee species.

These bees have a particularly fearsome reputation in the United States, where they were (accidentally) introduced from northern Europe sometime in the 1960s. Much as the Harlequin ladybird is outcompeting a number of UK ladybird species, so *A. manicatum* appears to be outcompeting the native American *A. maculosum*. Although I have seen bees being chased away or knocked to the ground by dominant male Wool Carders, I have never myself witnessed a fatality.

A week later and I am checking the lamb's ears again. I have been doing this every day. It is every bit as busy today as it was earlier in the week, but still I cannot see any Wool Carders, male or female. Unusually for solitary bees, the females of this species emerge from their natal nests before the males. I wonder if Juliet has seen a male or female. *Maybe I should wander up to Juliet's garden*, I think, *as they're clearly not interested in visiting our plot. Unless... Hang on a minute... Is it...?* I catch my breath and hold it until I'm sure. *It is!* A beautiful Wool Carder female is resting, completely still, on one of the lower leaves of the lamb's ears. Had I been standing just a few centimetres to either side, I wouldn't have seen her.

As though she has been waiting for my arrival, she shifts a bit, then proceeds to bestow me with one of my most treasured bee moments ever, as she begins to 'card' the silken hairs of her chosen leaf. I have seen dozens of Wool Carder bees before, but never witnessed one harvesting the fine, soft, downy hairs with which they line their nests. How thrilling it is to watch her at work. She is bent, almost double, pivoting herself on the leaf with her four back legs whilst she uses her front legs and her powerful mouthparts, or *mandibles*, to scrape the hairs into a little ball of fluff beneath her abdomen. I watch, entranced, relishing every single last second until she gathers up her little ball of fluff and flies home, her prize safely secured in her jaws.

Later, I mention this encounter to solitary bee expert Stuart Roberts, who asks me if I have read the great English naturalist Gilbert White's observations of this bee in the village of Selborne, East Hampshire, dated 11 July 1722. I hadn't, but I am intrigued. White writes,

> *There is a sort of wild bee frequenting the garden-campion. It is very pleasant to see with what address it strips off the pubes, running from the top to the bottom of a branch, and shaving it bare with all the dexterity of a hoop-shaver. When it has got a vast bundle, almost as large as itself, it flies away, holding it secure between its chin and its forelegs.*

Stuart suggests that the 'garden-campion' White refers to might be *Lychnis flos-jovis*, commonly known as the flower-of-Jove. What better excuse to order some seeds.

I spend the rest of the afternoon hovering around the lamb's ears, hoping for a repeat performance, but though I see a number of female Wool Carders collecting pollen, there is no more carding.

Of all the cavity-nesting solitary bees in the world, the larvae of this bee must surely be blessed with the most comfortable start in life. Other cavity-nesting species, such as masons and leaf-cutters, must rely on base materials such as mud, leaves, resin, flower petals, and pebbles. Only the larvae of Wool Carders hatch from their eggs into a nest of such softness. And she is an extremely skilled nest-maker, this bee. When she arrives back at her nest with her ball of fluff, she teases it apart with her sharp multi-toothed mandibles, then, using individual fibres, weaves together a finely wadded purse inside which she deposits a little lump of pollen, before laying a single egg and then sealing it with more wadding. The process by which she first gathers, and then manipulates, the plant hairs is known as *carding*.

In his book *Bramble-Bees and Others*, the nineteenth-century French naturalist Jean-Henri Fabre describes the nesting habits of *Anthidium* bees, which he calls 'cotton bees' and 'cotton-weavers', for the balls of fluff they harvest. So desperate was he to see, first-hand, how they made their little purses, that he tried to lure them into his study to build nests in bee boxes outfitted with glass tubes. He left the door from his garden into his study open throughout the summer, but 'not once did the Cotton-weavers... condescend to take up their quarters in the crystal palaces.'

No one writes quite so beautifully, or with more attention to detail, about the nest-making skills of this enchanting insect than Fabre:

> No bird's-nest, however deserving of our admiration, can vie in fineness of flock, in gracefulness of form, in delicacy of felting with this wonderful bag, which our fingers, even with the aid of tools, could hardly imitate, for all their dexterity. I abandon the attempt to understand how, with its little bales of cotton brought up one by one, the insect, no otherwise gifted than the kneaders of mud and the makers of leafy baskets, manages to felt what it has collected into a homogeneous whole and then to work the product into a thimble-shaped wallet. Its tools as a master-fuller are its legs and its mandibles, which are just like those possessed by the mortar-kneaders and Leaf-cutters; and yet, despite this similarity of outfit, what a vast difference in the results obtained!

I prefer books by this author, above all others, for my bedtime reading.

Reading Fabre has me wanting to find a Wool Carder female nesting in one of the bee boxes in our own garden, but alas this has yet to happen. My understanding is that Wool Carders like to nest up high, so our boxes are probably all in the wrong positions. I say this because I have recently seen a video someone took of wool carders nesting under old roof tiles.

Also, my friend Vivian Russell found a Wool Carder nest, in between two boxes of film slides in her upstairs studio last year. A female Wool Carder had clearly been flying in and out of her open skylight window for a few weeks during the nesting season without her knowledge, but it wasn't until months later, in November, that Vivian discovered the nest. Vivian is a photographer, and the images she took of the nest are exquisite. The nest cells look like tiny, felted wallets, exactly as described by Fabre in his book. And you could see right through the fine wadding to what was probably the cocoon inside. Curiously, each wallet appeared to have a little tapered end to it, but I do not know if this is normal, and if so, what the purpose of the tapering might be. Perhaps it was just where the female sealed the cell off.

Anyone else finding tiny, cotton-wool nests inside their home would probably have hoovered them up, or swept them into a bin, along with other miscellaneous pieces of fluff. But Vivian has for many years been planting her garden specifically to attract pollinators, and having noticed some Wool Carders visiting her *Stachys* earlier in the year, she guessed immediately what they must be. It would, she knew, be risky to leave them in a heated room indoors over winter, so she removed them carefully to a sheltered spot outside – an old wooden banana box where she keeps other solitary bee cocoons over winter. She makes sure to keep the lid slightly open, so the newly emerging bees can escape when they hatch out. All being well, the adults will emerge next year, along with the mason and leaf-cutter bees they share the box with. This Wool Carder female made a good choice when she decided to build her nest in Vivian's studio.

I usually try to avoid doing research on the internet just before I go to bed, but after seeing Vivian's photographs of the Wool Carder nest, I read paper after paper well into the early hours that night. I was already in awe of the Wool Carder female's craftsmanship, but as it turns out, even more goes into the preparation and construction of the nests than Fabre had made me aware of. I knew that they collect plant hairs to construct nests, and that they gather pollen, often from the same plants, to provision their brood cells. However, it seems the females visit some plants for yet another reason: to collect secretions that help to guard their brood against certain threats.

I have already mentioned that Wool Carder bees have creamy white hairs on their tarsi. Like some other species in the Megachilidae family to which they belong, Wool Carders use their specially adapted tarsal hairs to absorb secretions from the tiny hairs situated on the stems and buds of plants, including geraniums and snapdragons. However, whilst some bee species use floral oils and other plant secretions to help waterproof their nests, or mix them with pollen to feed their larvae, the Wool Carder bee saturates the finished cotton purse with the plant secretions she collects. Scientists believe chemicals in the secretions may help to deter parasites, therefore offering some extra protection to the brood.

Having done all she can to give her offspring the best possible start in life, the adult female, like most other solitary bee species, will sadly not live to see the fruits of her labour. She will never see her larvae construct further defences to protect themselves during pupation. Whereas other cavity-nesting bee larvae spin silken cocoons around themselves, Wool Carder larvae, according to Fabre, ingeniously apply a coating of their own faeces to the inside of their fragile little cotton brood cells. In photographs of the cotton purses, cut open after the bees have flown, the interior looks like a pale copper-coloured shell.

I am overjoyed when the Wool Carder bees finally appear on our allotment so that I might observe them myself. I have been quite concerned over the last few years that their numbers might be declining, but perhaps like many other creatures, they have good years and bad ones. Either way, it has been worth the wait. And if females are using our lamb's ears to collect the hairs to line their nests, there must be at least one male around. I must have simply missed him.

When I get home there is another email from Juliet. She has been watching a male 'roaming his territory', she writes. She saw him mating with more than one female, and she has also witnessed him 'charging a Red-tailed bumblebee and knocking it to the ground'. I am so very pleased for her – and looking forward sometime in the next few days to seeing this gladiator bee on our allotment, too.

I think next year I might ask Rob to attach some bee nesting boxes to the wall just under the eaves of our house, and I will plant lamb's ears and other *Stachys* in the bed directly underneath and, on Stuart's recommendation, white horehound and purple toadflax. I might even take a leaf out of the Reverend Gilbert White's book, and try some garden campion.

There are few lengths I would not go to in my efforts to see this species nesting in our garden.

CHAPTER 17

Time for Tea

Having spent as much time as I have over the last decade watching bees and other insects, I inevitably began to pay more attention to the flowers they were visiting as well. So a few years back, after taking a short wild flower identification course offered by the Species Recovery Trust, I decided to further develop my botanical detection skills and signed up for an online course with the Botanical Society of Britain & Ireland (BSBI). I had good intentions, and set aside time for study and fieldwork in order to complete the fourteen assignments in the allocated time. What I hadn't bargained on was my mother becoming ill.

I tried, against the odds, to soldier on with the course but failed to get past the fourth assignment. I simply did not have the mental or emotional capacity to continue. With a heavy heart, I gave up.

In retrospect, it wasn't just caring for my mother that prevented me from completing the course. When comparing the experience with previous courses I had attended, I realised that online courses are not ideal for me. I learn more within a group of people, face-to-face, when I can instantly bounce ideas around and build a rapport with others. And as I reflected on it, I realised that I also learn more when I am outside, in nature.

So I head back to the garden, to the allotment, hedgerows, woodlands, hills, and downs of North Dorset, in the hope that the plants themselves might become my teachers. I feel I might be coming full circle back to where I was as a child, before I was taught 'how to learn'. I will certainly keep my reference books (I love them), continue to look things up online, and carry on asking for help and advice on Twitter from bee lovers and experts alike. Mostly, though, I trust that the secrets and truths for which I am searching will make themselves known to me if I open my mind and listen carefully to my heart.

My friend Rachel Corby knows all about this. She listens to plants in ways few of us know how, or dare to try. Rachel's approach to gathering knowledge of plants and their secrets is akin to the manner in which our ancestors developed their cornucopia of plant lore and medicine. Indeed, many Indigenous peoples continue to identify uses for plants not merely through the oral traditions passed down by their elders, but also by communicating with the plants themselves, in the purest sense of the word.

Have you ever tried to communicate with a plant? Probably not. I hadn't either, until I met Rachel.

I first met Rachel at a small 'off-grid' gathering in Somerset where she was leading a medicinal plant walk. As I have always been fascinated by the therapeutic benefits of plants, this walk was at the top of my list of must-do's for the weekend. I was looking forward very much to learning more about the properties of our native hedgerow plants. I was not expecting to have my world view, or at least the way I viewed plants in general, turned upside down, but that is what happened.

As we meandered along the hedgerows, stopping every now and then so Rachel could introduce us to this plant or that, and talk to us a little about their properties and *personalities*, it quickly became evident, from the way Rachel spoke about the plants, that the knowledge she was sharing with us had not come from conventional books or courses on herbalism. It wasn't so much that her knowledge differed from the knowledge other herbalists have of plants and their uses, but rather that she had come by it in an entirely different way, from the plants themselves, perhaps. My interest was aroused, and as soon as I got back home after the gathering, I ordered Rachel's book *The Medicine Garden*.

What a treat it was to read this book. On a practical level, it contained all the information I needed to be able to identify which plant to pick – for instance, for a sore throat, sinusitis, or sunburn – as well as giving me detailed instructions on how to make basic preparations such as tinctures, salves, infusions, and syrups. But what most inspired me was Rachel's deep and intimate connection with the plants she writes and speaks about, from those that grow on her own back doorstep, lawn, flower border, and vegetable garden, through to those she has come across farther afield in hedgerows, woodlands, riversides, and mountains.

Rachel's teachings were unlike any I had ever come across before. They were a call back to nature, a reminder that we are in danger of growing so far apart from the connections some of our ancestors once had with, in Rachel's words, 'the conscious and sentient world [we] are embedded within', that if we are not careful, we may never find our way back home.

So inspired was I by Rachel's way of seeing plants, and the wisdom she receives from them, that I signed up for a weekend Communicating with Plants workshop at her home in Stroud to find out more. We were asked to bring a notebook, a pen, and some coloured pencils with us. I imagined my notebook might get filled with 'information' and botanical drawings of plants, but I could not have been further off the mark. I was aware from having met Rachel, and from reading her book, that her workshop might well take me slightly out of my comfort zone, and so it did, in more ways than I expected. But it is good to be taken out of your comfort zone every now and then; some of the most wonderful and memorable moments of my life have come at times when I have dared to step out of this zone and into the unknown.

There were four others on the course, and after we had introduced ourselves and settled in, Rachel invited us to drink a cup of 'mystery tea', which she poured from a teapot into little glass beakers. She asked us to drink the tea slowly and then jot down any thoughts and feelings that came to us.

The first thing I noticed was the taste. It was subtle, fresh, and 'green', but unlike anything I recognised. I felt an overwhelming sense of being 'supported', as though my burdens were being lifted off my shoulders. We compared notes with one another and were astounded to discover that much of what we had scribbled down was remarkably similar.

Rachel explained how, in the absence of her giving us any clues or information about the plant from which she had harvested these leaves, we had
all tapped into a deeper connection, or intuition – call it what you will, but,
fundamentally, a form of communication between the essence, energies, or
spirit of the plant and ourselves.

We repeated this exercise with another plant infusion, but this time Rachel
asked us to think specifically about whether, and how, we might benefit
health-wise from drinking it. The same thing happened – that is, everyone
jotted down very similar thoughts about its possible medicinal properties.

After a delicious lunch of homemade soup and salad, Rachel ushered
us into her garden. We were invited to find a dandelion plant and spend
as much time as we wanted, or needed, sitting quietly with it, examining
it, and noticing its scent, where it was growing, and other aspects of its
personality. Most of all, she said, we should concentrate on how our chosen
dandelion made us feel whilst we were sitting next to it. Did it make us feel
more awake? Tired? Happy? Sad? Did it perhaps bring back memories?

After spending this time communicating with the dandelion, we were
to sketch our plant and when, or if, we felt ready, ask it, *Do you have any
medicine to share with me?* Before we began, Rachel had suggested we offer
the plant some kind of gift – say, a few drops of water or a compliment – in
much the same way as you might bring a small gift to someone who has
invited you over for dinner. Above all, Rachel invited us to remain open-
hearted and open-minded during this dialogue with nature.

I sit down and begin, pouring a few drops of water onto the leaves of my
dandelion by way of introduction. Next, I say hello and thank the plant
for brightening up my day with its sunshine-yellow petals – not out loud,
but in my thoughts, with the express hope that the plant might 'hear' me.
Then I make myself comfortable on the grass and look – *really* look – at the
dandelion, as I never have before at any other plant.

At first I feel a little self-conscious. It's not as though I'm unused to sitting and watching something intently; after all, I do it all the time with
bees. But this feels altogether different. Plants are stationary, firmly rooted
in the ground. They don't pull away, or take fright and fly off, when you get
too close. They stay put. Unmoving. What was I supposed to see or hear?

It doesn't take long, though, before my mind and body become still and
I forget my awkwardness. I begin to look more closely at the dandelion and

notice its distinct parts: first its buds, then its flowers, then the individual petals, which, in turn, draw me down to the stems and leaves. I notice the shade my plant casts on the grass, the direction its flowers are facing, how bold it seems, sitting there, holding its ground in the middle of Rachel's lawn without a care in the world.

When I am ready, I pick up my pencils to draw, tentatively at first, but then, egged on by the dandelion, boldly. Its personality is wearing off on me. I am no artist, and my sketch is not accurate or to scale. But though I would not have thought it possible after having examined the plant so closely, I find myself seeing even more details whilst I am drawing than I had noticed whilst I was simply looking at it. I carefully shape the tips of each floret to show they are toothed rather than rounded; I try my best to capture the perfect curl at the tip of each bilobed stigma, the dents in each leaf, the hairs on each stalk. Fancy it having hairy stalks! I had never noticed hairs on dandelion stalks before. As I sketch away, I totally lose track of the time.

Drawing my dandelion is in itself deeply meditative, but after I stop drawing, I am surprised anew. I am just sitting with it now, not trying to force anything, not trying to make out its details or personality, when I begin to feel something more, something I can only describe as a kind of 'knowing' or 'awakening'. We do not, as a rule, spend time communicating with plants, so lack the vocabulary to describe such communications, or encounters, but I am sure some of our ancestors would recognise the sense of 'oneness' I am experiencing.

When I rouse myself from the moment, I feel no need to ask the dandelion if it has any medicine to share with me; it has already given me enough. I thank the plant from the bottom of my heart for the new awareness it has bestowed upon me and go into the house to join Rachel and the others.

Rachel had told us that there were many different levels and layers of plant communication, and that we didn't need to understand them all to feel their benefits; everything that we would hear was part of the medicine the plant has to share. Still, when those in our group shared what medicine their plant had for them, incredibly, they once again jotted down very similar things. The thoughts and feelings also mirrored those we had expressed in the morning, after we'd sampled Rachel's tea. It had been – *of course* – dandelion tea.

People come into your life at different times and for different reasons. Had I met someone like Rachel in my late teens or early twenties, I would have been extremely open to her ways, and my life might have followed an entirely different path. From a very young age, I had been interested in whole foods and natural remedies, and by the time I was in my early teens, I was drinking tea made with sage from the garden to soothe a sore throat and placing sprigs of lavender on my pillow to help me sleep during school exams.

By my mid-teens, I was spending hours in the kitchen brewing up concoctions for my mother and her friends to use on their skin, religiously following the recipes in Clare Maxwell-Hudson's *Natural Beauty Book*. I would carefully warm the water- and oil-based ingredients in separate pans, mixing my oils and waxes in a Pyrex bowl above a saucepan of boiling water so they never came into direct contact with the heat. Then I'd patiently add the oils, drop by drop, much as you do when you make mayonnaise, to the warmed orange flower or rose water. I dreamed of making a living from selling these homemade lotions and potions. (This was in the mid 1970s, shortly after VAT was introduced, and I am ashamed to admit that I boosted my income by adding 10 percent to all my prices, though I hadn't a clue what VAT was and didn't send the money on to the taxman.)

I tested all the recipes at home before selling them. It must have been pretty challenging for my mother, when she cleaned the house, to find dried-up remains of thinly sliced cucumbers and potatoes on the bathroom floor, and dollops of double cream mixed with runny honey in the basin. Not to mention the mashed avocado face masks – no longer avocado green, but sludgy brown and lumpy – on the edge of the bath. She was very tolerant, especially as she never understood or shared my interest in such things.

In fact, apart from my slightly older cousin Sarah (my teenage hero and life model), who lived hundreds of miles away in Suffolk, I don't recall any other members of my family being in the least bit interested in natural foods or remedies. So I cannot for the life of me work out why I was. Maybe I had been influenced by the books I read as a young child, or I had an innate instinct of some kind to avoid anything 'unnatural'. I don't know. Regardless, I nurtured an awareness of the importance of working 'with' nature rather than 'against' her, and even during those decades when I was otherwise disconnected from the natural world, I kept drinking sage tea and taking other 'hedgerow medicines'. This makes me feel a bit better about those lost years. I'm not sure why, but it does.

We each relate to the natural world in our own way. Why wouldn't we? After all, our relationships with other human beings take on so many forms that no two relationships are ever the same. If you think for a moment about the relationships you have with friends, family members, work colleagues, or lovers, you will struggle to find any two alike. Some will be deep and meaningful, some challenging, and some easy. There are relationships that tug at your heart, and others that stimulate your mind. Some are based on shared interests, others on material needs. Much the same is true of our various relationships with aspects of the natural world.

And whilst many of us view this planet as a wondrous, magical place to be loved, cherished, and cared for, others view it as nothing more than a 'resource', something to be managed, controlled, and exploited – then discarded when it no longer serves its purpose. I am saddened that so many of us no longer respect and honour the land, never mind the plants that feed us, and help and guide us on the road to recovery when we are ill. If only we could experience and embrace the sort of deep, sacred connections that our ancestors had with the land, the world would surely be a kinder and better place. I believe we *can* experience these connections, if we put our mind to it, although to do it, we need to engage not just our minds and intellect but, more important, our hearts and senses, too. If our hearts remain closed and protected, and our senses dulled, we will never be able to fully appreciate or be touched by the beauty and wonder around us.

These things cannot be forced. For me, it has helped to create pockets of time and space to see what might happen, to open myself to new realisations and understandings of the kind you cannot absorb from books. These are my 'pockets of stillness', time and space to *be* rather than to *do*. Creating these pockets is not so simple as it sounds, but, I think, it is essential. And as I found out by sitting in front of a dandelion on Rachel's lawn, notebook and pencil in hand, waiting for some flash of eureka, it is also worth waiting for these moments to come to you.

———

Seeing dandelions, loads of them, as I walk down through St James's Park in Shaftesbury this afternoon, reminds me of my encounter with the dandelion in Rachel's garden. They have such a bright and cheerful disposition, not to mention an air of defiance, for one so vilified. For such cheerful-looking plants, they seem to attract an awful lot of negative press. Why they are so maligned I do not know. As far as I am concerned, there is nothing more joyful than a field, lawn, bank, or roadside verge covered in dandelions. I am so happy that our council have not chopped or sprayed the dandelions in this park. Every year, they are absolutely covered in numerous species of bumble-bee, solitary bee, and hoverfly. I have noticed huge swathes of them growing on some of the verges between Shaftesbury and Cherry Orchard, too, and am enormously relieved to see that these have not been cut down either.

Roadside verges are not what they used to be. I remember walking, when I was little, along the sides of roads with wild flowers growing in abundance, almost as high as the hedges themselves. These days, verges (and hedges) are often cut or flailed to within an inch of their lives, which is a tragedy given their importance as *wildlife corridors*.

It is well known that losses of wild flower meadows and grasslands over the last century have caused great declines in wildlife. Less well known is the fact these losses and declines have been compounded by many of the remaining pockets of rare and beautiful flower-rich grassland becoming isolated through lack of adjoining wildlife corridors to link them up. Wildlife needs to be able to disperse and migrate freely between one habitat and another, not least to maintain genetic diversity. Without corridors to provide cover and food for them, they cannot move freely and safely between their natural habitats.

The term 'wildlife corridor' does not just mean verges, however; it refers to any linear feature in the landscape that can be used for migration or dispersal of wildlife – railway tracks, ditches, streams and riversides, field margins, and avenues of trees. And linking small, urban wildlife gardens to one another is every bit as important as linking the huge, fragmented wild flower–rich habitats that still remain in the open countryside. Fortunately, organisations such as River of Flowers, a social enterprise founded by Kathryn Lwin, and the charity Buglife, together with their partners, are beginning to tackle this problem by working with local people to create not just wild flower–rich areas but also corridors between these areas. The charity Plantlife is simultaneously raising awareness of the importance of restoring roadside verges to their former glory.

Still, of all our wildlife corridors, it is perhaps the hedgerows, when well managed, that are capable of supporting the greatest number of species, both flora and fauna. According to the Woodland Trust, some five hundred to six hundred plant species alone have been recorded in English hedgerows.

Nothing, and I mean nothing, lifts my own heart more than the sight of a mixed native hedgerow, together with the community of wild flowers, grasses, mosses, fungi, and lichens that live and thrive within its boundaries and extended root systems. Mixed hedgerows, with their cloaks of sundry greens and embellishments of hips, haws, berries, fruits, and nuts, constantly surprise with their ever-changing tapestry of colours and textures.

Even in midwinter, the hedgerow's predominantly grey-brown twigginess is interspersed with the glossy greens of holly and yew; the bright sulphur yellows and pale sage greens of different species of lichen; and blocks of crinkly copper-coloured beech, its leaves too stubborn to let go and fall, despite its neighbours having long since dropped and massed together in a rich, dark mulch around the roots. As well as feeding the roots, this deliciously rich leaf litter provides a valuable habitat for all manner of invertebrates, which, in turn, provide meals for others.

Mixed hedgerows are mostly made up of woody staples such as hawthorn, blackthorn, hazel, beech, and oak, but might also include patches of privet, spindle, guelder rose, holly, crab apple, buckthorn, wild plum, hornbeam, field maple, or yew. Where space permits, other trees might have been added to the mix for extra height and shade – ash, for instance, or white beam, rowan, and birch. Then there are the climbers: bramble, honeysuckle, ivy, and old man's beard (traveller's joy, my mother used to call it), weaving and stitching their way in and out, up, over, and through their hosts. There is no limit to the possibilities for delight you will find in a mixed hedgerow.

Hedgerows, by definition, include any herbaceous vegetation, ditch, bank, or tree within three metres of the centre of a hedge. The ditches alongside our hedgerows help expand the range of the hedgerow ecosystem, dramatically increasing its biodiversity by providing habitat and cover for frogs and toads, which use them as safe corridors in their search for ponds, and by creating the damp, shady conditions favoured by woodland plants such as spurge, wild garlic, and wood anemone; numerous ferns and mosses; and in summer, meadowsweet.

Meadowsweet (*Filipendula ulmaria*), with its frothy clouds of creamy yellow flowers, is perhaps one of the prettiest, and certainly the most fragrant, of hedgerow plants. I can't quite decide whether its scent reminds me more of honey or almonds; perhaps both, which is why it is so surprising when I crush it that it smells more like antiseptic.

There is a hedgerow on the outskirts of the village of Sedgehill with a huge, wide ditch running alongside its entire length. Here, meadowsweet grows in equal quantities with great willowherb (*Epilobium hirsutum*), both reaching heights of up to two metres or more. The creaminess of the meadowsweet provides a perfect foil for the splashes of fuchsia-pink willowherb in the weeks during which their flowering periods coincide. The meadowsweet flowers make a very delicious almond-flavoured tea, too.

There are many mixed hedgerows flourishing along the roads, lanes, and tracks of North Dorset, but my favourites by far are those which run along French Mill Lane, leading out from Shaftesbury towards Melbury and Compton Abbas. These hedgerows, in late spring, truly are a sight to behold. They remind me of the glorious Cornish hedgerows that sit high above the roads atop colourful banks of bluebells, red campion, stitchwort, yellow archangel, and more. I always feel sad when these flowers begin to fade, but the saving grace is that, as they do, the hawthorn begins to flower.

Hawthorn (*Crataegus*) is one of the mainstays, if not the mainstay, of our beautiful native hedgerows. During its flowering period, it leaves most of its hedgerow companions very much in the shade. At first, the tiny, round pearly buds hang in clusters amongst deeply lobed, pale green leaves, giving the impression of a liberal sprinkling of late spring hailstones. But when they are fully open, the branches and leaves are almost completely obscured by the flowers cascading in waterfalls down towards the ground till you feel sure the branches are about to break under their weight.

Each flower comprises five snowy-white, dish-shaped petals speckled with stamens, tipped with dark pink, which becomes darker, almost brown, within twenty-four hours of the flower opening. Standing proud at the centre of each flower is a pale yellow-green *stigma* (the female reproductive parts of a flower, which receive pollen). The individual flowers are quite exquisite but can easily be lost in the sheer mass of blooms that cover the tree or hedge.

It is not just the appearance of this May blossom that catches your attention, though; if you were to walk blindfold past this tree, you would notice it just as well by its highly scented flowers. 'Scented' implies an appealing, perfumed fragrance, but in actual fact, the scent can be rather overpowering – sickly sweet and carrion-like. Flies love it.

Hawthorn, though, is more than just a pretty face; its value to wildlife is quite remarkable. Over the years I have encountered numerous new (to me) species living, feeding, and sheltering amongst its tangled twigs and branches.

One of its most common and easily identifiable residents is the handsome green-and-red Hawthorn shieldbug (*Acanthosoma haemorrhoidale*), whose nymphs feed on the ripening berries. It is one of the largest shieldbugs in Britain and Ireland, and can be seen from April through to October, though it is camouflaged beautifully amongst the leaves (which are reddish coloured when they first appear) and the berries, so you need to look carefully to find it.

Then there are all the wonderfully named moths that feed and lay their eggs on the leaves, branches, and bark of this tree: Pear Leaf Blister, Small Eggar, Light Emerald, and Fruitlet Mining Tortrix, to mention just a few. Also there is the Hawthorn moth (*Scythropia crataegella*), a small, mainly whitish moth with dark red-brown markings.

As for bees and other floral visitors, pretty much anything that is flying during the month of May will readily take advantage of the nectar and creamy-coloured pollen provided by hawthorn flowers. Butterflies, moths, bumblebees, solitary bees, wasps, and hoverflies all visit it in the hedgerows. Honeybees, especially, are attracted in great numbers, for May is the brood-rearing season, and hawthorn provides an excellent source of pollen, rich in proteins, vitamins, and antioxidants. Thank goodness it attracts all these flying insects, because even though hawthorn does not rely completely on its insect visitors for pollination, fruit set would be significantly reduced without them.

Then come the berries, or haws, from which this tree derives its name. On a good year, the branches bend and bow under the weight of abundant scarlet-red berries, which hang in clusters from the now much darker, mature green foliage. These autumn berries never last long. No sooner have they ripened than they are seized and feasted upon by hungry blackbirds, thrushes, and other native birds, and small mammals such as dormice – so long as migrating waxwings, redwings, and fieldfares don't reach them first.

Alongside its attractiveness to invertebrates, many of which also shelter and make their homes in its deeply fissured bark, the cover hawthorn affords is also enormously important for birds and small mammals. Numerous bird species make their nests inside its dense, thorny crown, whilst also using the thick, tangled cover for shelter and protection. Sparrowhawks would never follow a bird into a hawthorn hedge, for fear of damaging their feathers on the sharp, prickly thorns, which makes it the perfect place for small birds to hide. Frogs and toads find shelter here, too, as do mammals such as hedgehogs and stoats, which use its tightly laid trunks and roots not only for shelter but as corridors for travel, too, connecting them to habitats up and down the country.

———

Since attending Rachel's workshop, I have wanted to spend more time getting to know specific trees, and though I kept thinking I might like to start with oak, ash, or birch, it has been hawthorn hedges that have drawn my attention whilst I have been out on my walks recently. Interestingly, for all the time I have spent walking the hedgerows, I have not been able to decide whether hawthorn's personality feels male or female to me. Not that it needs to be one or the other; I began to wonder about this only after looking into the folklore surrounding the tree. It was said in the past that hawthorn was known for its duality and sense of balance. With its delicate flowers and sharp, protective thorns, it seems to combine both male and female attributes.

In the past, people used to gather boughs of hawthorn blossom and weave it into wreaths and garlands to decorate their hearths on May Day. I was surprised to learn this, since 1 May is far too early for hawthorn to be flowering. However, it seems that prior to 1752, which is when we switched to the Gregorian calendar, May Day actually used to fall closer to what is today 12 May – around when hawthorn starts to come into bloom. How wonderful it is to know that knowledge of the 'old ways' is out there, waiting for us to tap into, if we so wish.

I would like to spend more time with hawthorn at different times of the year, especially during the winter, when it is quieter – no blossom, just thorns. It always looks so unwelcoming, forlorn even, but I think there might be more to winter hawthorn than meets the eye. I know not what I might discover, and prefer not to speculate, but I look forward to spending some time this winter, finding out.

CHAPTER 18

Evergreen

My friend's mum, Kirsten, tells me that Ivy bees are nesting in one of the fields just off Foyle Hill, the old road that leads out of Shaftesbury towards Cherry Orchard. She saw them when she walked there with her border terrier, Fred, earlier in the week.

'There are hundreds of them,' Kirsten says, before trying to explain to me their exact whereabouts. It's no good; there are far too many gates, fields, walls, and stiles for me to take in, so by the time she gets to the little track just beyond the church where I need to climb over the first of the stiles, I am quite lost in my head. Kirsten very kindly agrees to meet me the next day to show me herself where the bees are nesting.

Ivy bees (*Colletes hederae*) are relatively new to Britain. First spotted in Worth Matravers, Dorset, in 2001, they have since spread rapidly across the countryside, and have been recorded as far north as Saltburn-by-the-Sea, in North Yorkshire. In fact, this very day as I am writing about their spread, my friend Vivian Russell has been in touch to say she has identified a population on the other side of England, in Cumbria.

As its name suggests, this solitary bee species has a penchant for common ivy (*Hedera helix*), which flowers from early September through to
late October. Ivy bees will also visit plants such as autumn hawkbit and
dandelion for nectar, but they feed their larvae almost exclusively on ivy
pollen. They will forage for pollen on Asteraceae, too, if they emerge before
the ivy is in flower.

The ivy was late coming into flower in Shaftesbury this year, but is making up for lost time now and is alive with the sound of buzzing insects. With
common ivy, it is only the mature form that produces flowers and fruit. I
had previously assumed I was seeing two separate species of ivy on the walls
and trees where I see it growing, but in fact they are juvenile and mature
forms of the same plant. The difference between the two is quite marked.

In its juvenile form, ivy climbs or trails, whilst the adult form stands erect
and can be freestanding. The leaves are different, too. Those of the juvenile
form are deeply lobed, with three to five lobes, and have pale undersides,
whilst those of the mature form are more oval in appearance, almost heart-
shaped, and a deep, dark, glossy green in colour. The biggest difference is
the flowers on the mature ivy. The flower heads are small, around thirty
millimetres in diameter, shaped like spheres, and made of up to twenty-five
pale green florets. So tiny are the yellowish-green flower petals, that if it
weren't for the distinctively sweet, heady, honey scent of the nectar, you
might not notice them at all. With olfactory cues like this, and such easy
access to its pollen and nectar, it is no wonder so many insects are attracted
to this plant.

Ivy is much maligned and misunderstood, vilified by many as a strangler
and killer of trees, and chopped down, willy-nilly, by its accusers. There are
situations where ivy needs to be removed from trees, for instance, if it is
suspected of hiding structural problems, or perhaps where it is smothering
trees that support rare and scarce lichens. It can also occasionally cause the
collapse of smaller trees after strong winds, but this phenomenon is unusual
in healthy, large trees. These are exceptions rather than the rule, however,
and I cannot help wondering if people knew how incredibly important ivy
is for wildlife, whether they might think twice before so readily attacking it
with their axes and saws.

The Woodland Trust, whose very aim is to protect and campaign on
behalf of this country's woods, states: 'Ivy does not kill or damage trees, and

its presence doesn't indicate that a tree is unhealthy or create a tree safety issue in its own right.' Surely, coming from the Woodland Trust, this must dispel some of the myths.

Ivy truly is a shrub for all seasons, providing food and shelter for numerous species of wildlife throughout the year, especially during the winter months and in areas where there are few other evergreens. Birds such as wrens, treecreepers, blackbirds, and spotted flycatchers frequently nest in and behind its foliage, as do greenfinches (*Chloris chloris*).

One of my Twitter friends, Vic Savery, tells me he remembers as a child finding no fewer than half a dozen greenfinch nests in a small line of trees covered in ivy. Sadly, this beautiful bird has undergone a steep decline in recent years. If and when its numbers increase again, I will be sure to watch the ivy during its nesting season.

As a food source, ivy caters for the larvae of certain butterflies, such as the Holly Blue (*Celastrina argiolus*), and also for a number of moth caterpillars, in the spring and summer, but it is in the autumn that ivy comes into its own, its unassuming flowers opening to reveal an abundance of rewards for visiting insects. Myriad insects feast and depend upon this plant's pollen and nectar: Ivy bees, honeybees, bumblebees, wasps, flies (especially hoverflies), hornets, butterflies, and others whose names I do not know. Without ivy, they would struggle to find enough food to prepare themselves for the hard winter months ahead. Then, after the flowers, comes the fruit, in the form of little black berries whose high fat content provides valuable and much-needed winter nutrition for blackbirds, thrushes, redwings, robins, wood pigeons, and others.

Over the last few weeks, since the ivy has been in flower, I have seen large numbers of male and, more recently, female Ivy bees foraging in and around town. So far, I know of only one large nesting aggregation in Shaftesbury, in a field on the town's southern outskirts, belonging to my friends Sue and Angela of the Shaftesbury Tree Group. There will be many more, but frustratingly I have not yet found them.

It is hard to miss an Ivy bee aggregation when you walk past one, especially at the beginning of their flight season, when the males are on the wing. The males' behaviour is quite frantic, almost frenzied, as they fly back and forth, up and down, close to the ground above the nesting site in search of emerging females to mate with.

Indeed, I receive almost as many emails and phone calls in the autumn, from people worried about swarms of Ivy bees, as I do in the summer about Tree bumblebee swarms. However, there is no need for my enquirers to be concerned. The male bees are not equipped with a sting, and to be stung by a female Ivy bee, you would need to pick her up and handle her roughly, which, of course, I know you would not do. So, Ivy bees are perfectly safe to watch close up.

Ivy bees nest in individual burrows alongside others of their kind, usually in loose or sandy soil. I have found them nesting in numerous different habitats, including roadside verges, lawns, fields grazed by livestock, and coastal paths – always south, west, or east-facing, and located in areas where vegetation is sparse. The numbers of individual burrows can range from a few hundred to many thousands.

Given the right ground conditions, room to expand, and a plentiful supply of nearby ivy, the bees' numbers can increase rapidly. I came across a site in Cornwall, along the top of a sandy cliff on the South West Coast Path, where I estimated there must have been tens of thousands of nests spread over a vast area of short grass and bare earth, and this, by all accounts, is not unusual.

It is mid-morning when Kirsten takes me to see her Ivy bees, and when we arrive, their nests are mostly in the shade. There are bees flying, but nowhere near as many, she tells me, as there were the previous afternoon. Solitary bees are always more active when it is bright and sunny, so I resolve to come back later in the day, when the site will be fully in the sun.

Kirsten drops me off at our allotment. I ought to do a bit of weeding, but my mind is too full of Ivy bees, so I pick a few sprigs of fresh mint and head back home. Once home, I gather together the gubbins I need for my afternoon field trip and check my camera. Its batteries need charging, as do mine. Time for a cup of fresh mint tea. The fragrance released by the mint as I pour the boiled water over the leaves is intoxicating, and the taste rather exotic, especially if you add a slice of lemon, which I do.

Refreshed, and raring to go, I pick up my rucksack and make my way slowly back down to the Ivy bee field. My walk takes me through some of my favourite parts of Shaftesbury: along Pine Walk, where after dark we often hear, and sometimes see, tawny owls; then down St John's Hill to Raspberry Lane, where I cut through to St James's Church. It is here, between Raspberry Lane and the church, that I become aware of an unusual amount of flying insects. Something about the ungainly and erratic way the insects are flying, bumping first into me and then into each other, makes me stop awhile to see what they are.

They are ladybirds, and they are landing in their hundreds, perhaps thousands, on the stone walls that surround the Old Rectory gardens. They crawl this way and that, occasionally around in circles, but mostly upwards, investigating the cracks and crevices as they go. They appear to be on some kind of mission, but it all seems slightly chaotic. When I look closer, I see they are Harlequin ladybirds (*Harmonia axyridis*). Apart from the fact that Harlequin ladybirds are all quite large – between seven and eight millimetres round – you might be forgiven for thinking you were looking at numerous different species. The most common forms in Britain and Ireland are orange with fifteen to twenty-one small black spots, and black with two or four orange or red spots, but they can also be red or yellow, with varying numbers of black spots. No wonder they are called Harlequins.

Harlequins are a non-native, invasive species. Originally from Asia, they were imported to the United States in the early 1980s, and later to the

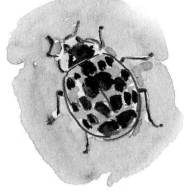

European mainland, and have since found their way across the Channel, probably on imported plants. Other alien animal species in the United Kingdom include grey squirrels, muntjac deer, a number of different crayfish species, American mink, and ring-necked parakeets, but, despite being here longer, most of these species are advancing at a far slower rate than Harlequin ladybirds.

In fact, the ladybirds have spread so rapidly that, according to Prof Helen Roy, who runs the UK Ladybird Survey, they are now the fastest-spreading alien species on record in the country. When non-native species establish a foothold in an area, they can cause great losses to local biodiversity, by spreading disease, outcompeting native species for food and other resources, and destroying habitat.

I have heard some gardeners say they dislike Ivy bees because they believe these ground-nesting bees are 'taking over their lawns'. I have also recently met a beekeeper who said he very much disliked Ivy bees, but not for this reason; he dislikes them (he said) because they take so much nectar from the ivy, thereby competing with his honeybees. He reported that since Ivy bees have arrived in his area, the amount of honey his bees produce over autumn has fallen enormously. Not only is he unable to draw off a large surplus the following year, but he told me that without the ivy nectar to boost their stores, they now struggle to get through the winter.

I understand, from this gentleman's personal perspective, why he dislikes Ivy bees, but it got me thinking: If the arrival of a few thousand Ivy bees can have this effect, so quickly, on existing populations of locally kept honeybees, can you just imagine the consequences of introducing hundreds of thousands of commercially raised honeybees to an area where there were previously only a few wild honeybee colonies and small-scale managed beehives? The effects on local native wild bee populations could be devastating.

It is important to make a distinction between those species that are impacting negatively on our native species, and those that are not. Tree bumblebees and Ivy bees, after all, are also newcomers to our shores. The spread of these two bee species is being carefully monitored by scientists, but neither appears to be competing with our native species. The Ivy bee for sure, and probably the Tree bumblebee, have both arrived naturally, by way of increasing the ranges of their existing territories from northern Europe.

Harlequin ladybirds, on the other hand, were definitely introduced by humans, and their presence here is, without doubt, causing huge declines in some of our once common native ladybird species. These invasive insects have voracious appetites, which is exactly why they were introduced to America and Europe in the first place; they were supposed to control aphids in crops. Unfortunately, their insatiable appetite for aphids means they are now outcompeting many of our native species, most notably the Two-spot ladybird (*Adalia bipunctata*), which shares the same ecological niche. It is no coincidence that populations of native Two-spots have fallen by 44 percent since the arrival of Harlequins.

The ladybirds on Raspberry Lane are looking for somewhere to hibernate, and the Old Rectory walls are clearly attractive to them. I take some photographs and continue on my way. As it turns out, I am not the only one to see hundreds of Harlequin ladybirds today. The long, warm summer this year appears to have resulted in record numbers of this species, which are now searching for places to hibernate over winter. Although those I saw today were outside, Harlequins are also very attracted to houses, often clustering together for warmth in the corners of window frames or ceilings, as they prepare for their long winter sleep.

The tabloids are typically full of scare-mongering stories about ladybird invasions at this time of year. The worst this year is in *The Sun*, running under the headline 'KILLER BUGS: Invasion of Cannibal Ladybirds Carrying STIs Wreaks Havoc in Homes across the UK'. Headlines like this add to the already irrational fear that many of us have of insects. If it's not invading ladybirds, it's infesting spiders, or infiltrating giant hornets, but they all end in the same result: people panicking so much they start indiscriminately killing beautiful native ladybirds, spiders, and hornets. We do need to be alert, as some of these non-native species do indeed pose a threat to native species, but I wish more care would be taken when they are reported. It feels, sometimes, as though we are at war with wildlife.

It is still early afternoon when I reach the Ivy bee field. As I climb over the gate, my attention turns upwards to the rasping 'caws' of a pair of crows mobbing a buzzard that has strayed into their patch. The buzzard is calling, too, her plaintive mewing a little shorter, perhaps more irritated, than usual. What a kerfuffle! I keep walking, but so distracted am I by the aerial acrobatics of the buzzard, as she swoops and swerves to escape the crows, that I walk straight past the Ivy bee aggregation.

The buzzard has dipped over the brow of the hill now, and the crows have dispersed, so I turn my attention back towards the ground. The field is dotted with devil's-bit scabious, vivid violet-blue buttons that, before I get my eye in, I mistake momentarily for butterflies. Blue is not a colour I associate with autumn, and the bright dots seem at odds with the rich reds, browns, and oranges of the surrounding trees and hedgerows.

I notice some of the scabious are growing clumped together, whilst others stand apart and alone. I like to think of these loners as pioneers, testing the edges of their territory, whilst the clump lovers play safe and stick to the tried and tested 'damp' areas in which they have already established a foothold. In actual fact, devil's-bit scabious naturally spreads out quite sparsely, so there is always more space between each of these plants than there is, for instance, between knapweed or yellow rattle.

I scan the field to see how far the scabious have spread from the little hub beneath my feet. They appear at first to be fairly localised, but when I climb the bank to get a better view, I can see they are beginning to colonise the entire field. I have seen meadows full of devil's-bit scabious in the past, and wonder how many years it will take for them to fill this field. But is this field, in fact, a meadow?

I find myself wondering what makes an ideal habitat for this scabious. We had a long, extremely hot summer, so this particular field – or, as I have mentally upgraded it, meadow – is currently drier than it would usually be in October. I am aware that devil's-bit has a preference for damp habitats, so it would be interesting to come back in wetter weather to see how the plants map to the various degrees of ground sogginess. They might make a good indicator of damp habitats.

I clocked on my way that devil's-bit scabious does not grow at all in the higher-up field, on the other side of the road, so my guess is that, when I return later in the year, the higher-up field will prove to be drier than this one. How marvellous it is that the simple act of taking time to sit with and observe a flower, in a field or meadow, can lead one to hypothesise about such things.

I dig out my pencil and notebook to make a few sketches. Although I had to abandon the online BSBI plant-identification course I enrolled in a few years ago, I did, before I left, learn a little about what to look for when trying to identify a plant. Just as I had with bees, my way of identifying wild flowers prior to the course had been to skim through a pocket handbook looking for an image of something that matched whatever it was I had in front of me. I still resort to this means of identification when I am stuck, but now I know to look very closely at every single part of the plant, taking in details of the structure, texture, and shape of the leaves, stems, sepals, and petals. It is quite incredible what you notice when you do this.

I look beyond the patch of scabious in which I am sitting. Those flowers that are growing in areas that have been heavily grazed by sheep and cattle appear a little more stunted than the others, which stand proud of the surrounding grasses on stems up to fifty centimetres high. I would like to investigate the flower heads, but they are bending and bowing too much in the breeze. I hold one steady by the stem so I can examine it properly with my field magnifying glass, which enables me to see and notice things that the naked eye would miss. I hadn't known such a tool existed prior to attending a beginners' weekend wild flower–identification course, but wouldn't be without mine now.

As I look through the magnifying glass, the first thing I notice is the flowers, which are fully open and far more complex in structure than I had previously realised. Each one is made up of numerous, tiny, four-petalled florets. From some of these, protrude pale lilac stamens supporting tiny pink *anthers*, which hold the pollen. On the smaller flowers, there are slightly thicker, flat disc–shaped *stigmas*, which receive the pollen. It appears the larger flowers might be male, whilst the smaller ones are female. None of this would I have seen without my magnifying glass. Nor would I have noticed the stems, which are round, dark reddish in colour, and covered in soft, silver-green hairs. The leaves, which are mostly at ground level or just below the top of the grass, are long and slender. Those at ground level are toothed, but farther up, their edges are smoother. I write all this down in my notebook and make some rough sketches of the leaves and the flower heads. One day, I would love to pick up the wild flower–identification course where I left off, or start it again from scratch.

I am brought back abruptly from my musings when a Common Carder bee (*Bombus pascuorum*) lands right on top of the very flower whose stalk I was next about to take hold of. The whole flower bends over under her weight. She is not the only pollinator I have seen visiting these flowers today. I noticed a number of hoverflies earlier, and Small Copper butterflies (*Lycaena phlaeas*), their dazzling, metallic forewings mirroring the colours of the leaves falling from the beech by the gate. This once common butterfly was in steep decline a couple years ago, but if the numbers I have seen this year are anything to go by, it would appear to be making a comeback.

Also, I am fairly sure that devil's-bit scabious is the larval food of the rare Marsh Fritillary butterfly (*Euphydryas aurinia*), but if this species is present in this meadow, I have not seen it – at least, I do not think I have. I must admit I would not be sure of its identification if I did see one. This is why I bring my camera on trips like this.

The Common Carder has extracted all the nectar she can from this particular patch of flowers. She flies up, and on, and I, too, must get up and on. Much as I would love to sit here a while longer, the sun is getting lower in the sky, and the Ivy bees will soon be heading back to their nests. I don't want to miss them.

Before walking back to the nest site, I do a quick once round the perimeter of the meadow to search for ivy. It is conspicuous by its absence. There are hips, haws, and berries in abundance, so plenty of winter forage for birds, but the Ivy bees must be travelling farther afield for their food. I come full circle to the bottom of the field, and am delighted to find the nest site is larger than I remembered – much larger. And it is buzzing with activity.

My plan is to take photographs and notes, but for the time being I am content just to lie on the grass and watch. I have no idea what time it is, nor do I care. Today, I am an Ivy bee watcher, and so long as the sun continues to shine on this part of the meadow, time matters not. I have been lying by the Ivy bee bank, not thinking, just watching, for a good ten minutes before it occurs to me that there are no male bees here.

Their absence explains why I walked right past the nest when I first arrived. The way Kirsten had described the 'swarms' of bees around this site suggests males were still around in large numbers earlier in the week. I must just have missed them. Apparently, female Ivy bees give off a pheromone, until they have mated, that keeps males close to the nesting site, so, like most other solitary ground-nesting species, these males disperse only after all the resident females have emerged and mated.

This nesting site has clearly been established for a good few years and is thriving. I am not surprised. The field is on a slope, and sheep have, over time, created a bit of a south-facing dugout, backed by an exposed, vertical bank that follows the contours of the land. I cannot see it all from where I am lying now, but the bank stretches round the corner for at least another fifteen metres and is, in some places, up to a couple of metres high. A south-facing bare earth bank – if you were an Ivy bee, this is exactly the kind of location you would choose to make your home.

The whole bank is pitted with small holes, each around eight millimetres in diameter. Some have been made by this year's emerging males and females, others by newly mated female Ivy bees burrowing in to make their own nests. I know from the BWARS website that the nesting biology of this species has been studied in some detail in Germany, where this species was described as new to science only as recently as 1993. The scientists write that each of the four tunnels they examined 'ran between 7–12 cm horizontally into a steep face, before turning downwards', and that 'groups of up to four cells, branched directly off the main vertical burrow, were located at a depth of 30–45 cm'. How marvellous to think that this, and other even more complex bee architecture, exists under the earth, (mostly) unseen by human eyes.

As the sun becomes brighter and warmer, the activity around the nest site increases. Some bees are bringing pollen back to provision new nest cells before laying their eggs; some are resting, facing outwards, just inside the entrances to their burrows; others are busily excavating new burrows. Old nests are being utilised, too.

There is so much going on, I hardly know which way to look. The newly created burrows are obvious from the little piles of fine soil spilling out where their constructors have been tunnelling, and the ground beneath the cliff face is covered in such deposits. These 'landslides' must make it extremely challenging for the bees that have chosen to make their nests in the ground beneath the cliff, for each time they return home they have to contend with the fact that their nest entrances have been covered up with fine tilth from the nests above.

It is fascinating to watch these bees searching for their own particular nest entrance. They know their nest is here somewhere, so they land, search, dig, fly upwards, zigzag above the ground to reorientate themselves again, land again, search again, dig again – and so it goes on until they finally succeed in locating the buried nest entrance. It must be incredibly challenging, having taken note of landmarks large and small, to find your nest entrance covered up on your return.

Having watched these and other ground-nesting species searching for entrances, I conclude that they must rely on visual cues. Were there some kind of pheromone to help them locate their own nests, I believe it would be easier for them. The bees that seem to find their nest the quickest are those that have entrances at the edges of the site, or close to vegetation of some kind or another – in short, where visual cues are clearest.

Those whose nests are in the centre of large aggregations, amongst dozens of other seemingly identical entrances, appear to spend more time searching for the right burrow before crawling in and depositing their pollen. Either way, they never give up, and I am full of admiration for their determination.

It is interesting to note the difference between the females that have newly emerged, and those that have clearly been flying for a few weeks. The former are pristine, their abdominal bands a deep burnt orange and black striped velvet, and their thoraces densely covered with ginger hairs. Those that have been around longer look dull and bald, their orange abdominal bands faded to pale yellow, their thoracic hair all but gone, and their wings noticeably tattered. This is to be expected, when you consider the hundreds of trips they make in and out of their narrow tunnels during their short lives on the wing.

One bee, in particular, has caught my eye. She arrived back from a foraging trip at least five minutes ago, the hairs on her legs laden with bright yellow ivy pollen, and flew straight to the entrance of her nest. For some reason that I cannot fathom, she seems intent on digging a larger entrance to her burrow. She has lost some of her pollen already, and will lose the rest, if she continues like this. Her behaviour is most odd. She would surely not have brought all this pollen back had she not already dug out a good, usable tunnel and created chambers in which to deposit it. I move in to get a better view.

There seems to be an obstacle, something stuck perhaps, blocking the entrance of her burrow. Maybe one of the bees nesting in the cliff face above has dislodged a small pebble and it has fallen into her nest? Her behaviour is becoming quite urgent now, almost obsessive. She tries to dislodge the obstacle by approaching it from different angles; a moment ago, she was squeezing her head and front legs inside the burrow, and now she is bent completely double. It looks like she is trying to grasp whatever it is that is bothering her with her mouthparts, tugging, heaving, pulling, but to no avail. She must be exhausted.

I have been watching her for well over twenty minutes when suddenly, without warning, she frees whatever it is and comes tumbling back down the bank. Or, I should say, *they* come tumbling back down the bank, for the obstacle, it turns out, is not a pebble but a second female Ivy bee! The two bees are locked together, in a yin–yang ball of wings, eyes, abdomens, and thoraces. It is hard to separate the two visually until she who I presume is the rightful owner of the nest – that is, the bee who just came back laden with pollen – breaks free and half scrambles, half flies back to the burrow, and disappears inside. At last, she can deposit what is left of her pollen load and provide for her brood.

But it is now time to go home. The sky has clouded over, there is a chill in the air, and I am hungry. I pack up my rucksack and wander back up the hill, stopping by the ladybird wall to see what is happening, but there are very few ladybirds left to see. They have disappeared into the cracks and crevices to sit out the winter. The Ivy bees will soon be gone, too, as will most of the other insects the ivy has supported during its flowering season. Some species will hibernate in their adult form, whilst others will fade away and die, leaving the entire future of their kind vested in the offspring they have left behind beneath the ground, or in old walls, rotten wood, roof spaces, plant stems, and so on.

I am so glad to have spent this day with the Ivy bees. There cannot be many more bright, sunny days like this left to come before winter. The bee season is coming to an end, which means that I am unlikely to see many bees again until next spring. I will miss them. I always do. But just like seasonal fruits and vegetables, their eagerly awaited return will be all the more special for their having been absent so long.

———

Postscript: I managed to get a couple of quick shots of the two Ivy bee females as they tumbled out of the nest. Later, when I crop and examine the photos, I can see that the pollen-laden bee did indeed pull the squatter out using her mandibles. I am impressed. Of course, I have no way of knowing which bee the nest really belonged to, but I *am* sure the returning forager believed it to be hers.

CHAPTER 19

Amongst the Snowdrops

It's 17 February and I have just seen my first Hairy-footed Flower bee of the year. She is foraging amongst the snowdrops in Diana's garden, and I can barely contain my excitement. Her small, black, round furry body emits the high-pitched buzz so typical of this species. It was this buzz which first alerted me to her presence in the flower bed. I watch her darting from flower to flower, her long, pointed tongue extended as she sups the nectar provided by these earliest of blooms, and I am smitten, all over again, by this charming little bee.

I cannot believe I don't have my camera with me. I have never before seen a Hairy-footed Flower bee foraging on a snowdrop. Usually, they are found on plants like lungwort, dwarf comfrey, wallflower, and red dead-nettle, not, to my knowledge, snowdrops. I would love to have a photograph to accompany my record when I submit it to the Bees, Wasps and Ants Recording Society (BWARS) later today. I wonder if mine will be the first sighting of the year…

The Hairy-footed Flower bee's scientific name, *Anthophora plumipes*, sounds to me almost as beautiful as her common name, though the *plumipes* part, meaning 'feather-footed', only really applies to the male of the species.

The male's middle legs are slightly elongated and adorned with long, soft, feathery hairs, which he uses to transfer secretions from his abdominal glands to the female's antennae during mating. It looks like he's covering her eyes with his legs whilst he does this. I have no idea what might be the significance of this transfer of secretions; this is yet another of the many mysteries I have still to unravel in my ongoing quest to understand more about the fascinating and beguiling world of insects.

Unusual amongst British bee species, male and female Hairy-footed Flower bees are markedly different to each other in appearance, although both are fairly easy to recognise and identify in their own right, even for complete beginners. Hairy-footed Flower females have jet-black bodies, with deep-orange pollen brushes on their hind legs, upon which they collect pollen. When fully loaded, these hairs are usually obscured, so might appear to be a different colour. Males of the species are gorgeously ginger in colour, their golden brown fading to a paler greyish brown after a few weeks exposed to the sun. They have pale yellow boxlike faces that look almost like snouts. And, of course, they have those extremely beautiful, and feathery, hairy, legs.

Although Hairy-footed Flower bees are a solitary species, they are often mistakenly identified as bumblebees because of their rotund, hairy bodies. However, you can be fairly sure it's a Hairy-footed Flower bee in early spring, when they first emerge, for the only bumblebees you are likely to see around at that time of year are the enormous queens who have just come out of hibernation. Compared with these huge bumblebee queens, Hairy-footed Flower bees are actually quite tiddly.

Colour, shape, and size aside, the easiest way to recognise Hairy-footed Flower bees – and to tell them apart from other bees – is undoubtedly by their behaviour. No other species of bee (apart from other related flower bee species, and, in some ways, Wool Carder bees) behaves, forages, or sounds quite like the Hairy-footed Flower bee. Zipping back and forth from flower to flower, with such speed and purpose that you can barely keep your eyes on them, then hovering for milliseconds in mid-air like hummingbirds as they probe for nectar and pollen, their behaviour really is most distinctive, and at times, quite un-bee-like. Add to this their distinctive, high-pitched buzz, and the male's territorial tendencies, and there is no mistaking a Hairy-footed Flower bee when you meet one.

Having said all this, as I watch my little flower bee climb carefully up inside each individual snowdrop, it slowly dawns on me that her behaviour is not quite so frantic as I have come to expect; in fact, she appears almost dozy. *She must have emerged this very morning*, I reason. *No wonder she seems sluggish*. I think again of my camera, sitting uselessly at home. I have spent days, if not weeks, stalking this species over the years, trying to get a good photograph, but they are usually flying too manically for me to get a shot in focus. Not this bee. She would have made a perfect subject for a portrait today, had I only been prepared.

I wonder, as I watch her, where she has come from. Though biologically solitary in their nesting habits, Hairy-footed Flower bees are actually quite gregarious and like to make their nests alongside each other, in soft coastal cliffs or bare, compacted ground, though more often than not, in old stone or cob walls, like the Bee Wall in Shaftesbury. I look around. There are plenty of old stone and cob walls in and around Diana's garden, as well as in the neighbouring farmyard. I expect there is probably a healthy population of Hairy-footed Flower bees nesting in the vicinity, just as there are at the Bee Wall in Shaftesbury. I must ask Rob to keep an eye out for them whilst he's gardening.

Hairy-footed Flower bees are one of a number of bee species that are generically referred to as 'mason' or 'mortar' bees. This is because Hairy-footed Flower females excavate short tunnels to reach the protective cavities inside the cliff, wall, or earth they have chosen for a nest. Once inside the wall, the female fashions individual, pitcher-shaped cells using the surrounding material – whether it be soft mortar, cob, or sandstone. Inside each cell, she deposits pollen gathered over numerous foraging trips, and then lays her eggs. She seals these cells, and the entrances to the tunnels, with more of her mortaring material.

When building their nests in walls, these bees burrow only into old mortar that has already been softened by the vagaries of time. Beyond the Bee Wall and other old stone walls around Shaftesbury, I have also seen Hairy-footed Flower bees in huge numbers on the gable ends of an ancient cob barn in northern France. All these walls will probably long outlive the owners of the buildings, and maybe even the aggregations of bees, though I do understand why owners of old buildings are concerned for the structure of their walls when there are large aggregations of bees nesting inside them.

I would love to have Hairy-footed Flower bees nest in our garden, but unfortunately the old stone wall in our garden is north-facing. However, where there is a will there can be a way, as has been proved by John Walters, who built his own cob bricks in the garden of his home on the edge of Dartmoor National Park. John's bricks have proved irresistible to the local population of Hairy-footed Flower bees, so Rob and I have decided to take a leaf out of his book, and plan to make a few cob bricks of our own this spring in the hope that we, too, might attract a passing female or two. I can't help wondering if perhaps homeowners who are worried about bees using their old walls could contain the problem by placing a few sacrificial cob bricks alongside the walls.

I bring my attention back to the flower bee foraging on Diana's snow-drops. She is resting now, on a leaf, so I can examine her better. She is perfect. But as I watch her cleaning her antennae with her front legs, my excitement turns to concern when I realise exactly *how* early it is in the season for this bee to be flying. My first Hairy-footed Flower bee sighting last year was well into March, and that was a male. Females usually emerge a couple weeks after the males. I do a quick calculation in my head. This female has woken up at least four weeks before she should have.

Today the sun is bright and the temperature has reached an amazing 9.2 degrees Celsius, but according to BBC Weather, there is a frost due tonight, and very little sunshine is forecast for the rest of the week. This does not bode well for my little flower bee.

———

20 February. I have seen neither hide nor tail of my Hairy-footed Flower female, nor any others of her kind, since I watched her on the snowdrops a few days ago. I know it in my heart, but it is difficult, nonetheless, to state it.

She emerged too soon.

Everything is so out of kilter these days, and the more tuned in you are to what is going on around you, the more obvious this becomes. You don't need to read or hear about the changes in the planet on the news; you can see and sense them for yourself.

An increasing number of pollinator species are beginning to emerge earlier or later than they used to; birds are pairing off and starting to nest when they shouldn't; and plants that used to flower in February are now flowering in November, or vice versa. There are many other examples, far too numerous to list, of changes in wildlife and plant behaviour. Many of us have grown aware of these changes, just as we have become aware of the changes in weather patterns. Worryingly, though, we are now beginning to *expect* rather than be surprised by the unexpected, and this, in turn, has led us to become more accepting of situations that should actually be causing us a great deal of concern.

We can shake our heads and talk about how different things were in the halcyon days of yore, but were the treasured landscapes of our childhoods really so very perfect? Compared with what we are witnessing today, they probably were, but our grandparents, having witnessed the massive changes that came about after the mechanisation of agriculture in the early and mid twentieth century, would most likely think differently. Whilst we hark back to the days when our gardens were full of butterflies, and we had to stop to wipe our car windscreens clear of flying insects, our grandparents' childhoods would have been enriched by the sight of majestic English elms and swathes of flower-rich grasslands, as well as the sounds of corncrakes, curlews, and turtle doves. These sights and sounds would have been as familiar to our grandparents as were the clouds of garden butterflies, and the poor squished insects on our windscreens, were to us. This phenomenon is known as *shifting baseline syndrome*.

Shifting baselines cause human beings' perceptions of what is 'normal' to change over time. Due to our short life spans and flawed memories, our baselines – that is, our collective and individual perceived knowledge of the world around us – shift with each successive generation. From a conservation point of view, we are measuring the degradation of the natural world based on childhood memories of what we assume was the norm, and more than that, of what we hold dear as a perfect state of nature. It is often this picture of perfection we strive to return to. However, what we saw as perfect twenty, thirty, forty, or fifty years ago, our ancestors would have seen as unrecognisably degraded. Our children, in turn, will view what *we* see as degraded as perfectly natural.

Change, per se, is neither inherently good nor bad. It is simply a part of life. How often have you found yourself sitting in a given moment wishing it would last forever, whilst simultaneously knowing it cannot? Or, conversely, had someone tell you whilst you are in a state of deep sadness, anguish, or distress that 'time heals', or that 'this too will pass'? I can think of many such moments. But there is change, and there is *change*. The changes in the natural world that we are currently witnessing are of such magnitude that we can barely comprehend, let alone assimilate, their significance and probable consequences. What will it take for the *balance* to finally tip from these changes causing mild concern to engendering unilateral alarm?

Many of us *are* concerned, some of us *are* alarmed, but the vast majority are merely aware, to some extent or another, that something is not quite right. Yes, there has been a shift in recent years in attitudes towards plastic waste, disposable products, composting, animal welfare, and the planetary impacts caused by overconsumption of meat and the shipment of goods thousands upon thousands of miles. But despite these shifts, some of them huge, humanity still appears to be oblivious to the true magnitude and urgency of these issues. And others, such as soil degradation, habitat loss, deforestation, pesticide contamination, loss of biodiversity, and the elephant in the room – climate change – are barely registering any changes in how we live on, and with, our planet.

Nature is in a constant state of flux, forever flowing this way and that. Individual species come and go, as do ice ages and interglacial periods. Landscapes change, sea levels rise and fall, and continents drift. But the changes in climate that we are currently witnessing are extremely unusual compared with those the planet has previously undergone.

Using evidence gleaned from tree rings, glacial ice, sedimentary rocks, and coral reefs, scientists have been able to build up a picture of average global temperatures and changes in the climate stretching back at least 800,000 years. Over this time, the Earth has experienced many extreme changes, but the rate of warming over the last century has been around ten times faster than previous rates, and this is predicted to double over the coming century. In the past two million years, average global surface temperatures have rarely approached the levels they are projected to reach over the next few decades. These facts are alarming, and the time to take action is now.

Climate change is already causing greater fluctuations in temperature, as well as excessive rainfall – resulting in flooding – one month, followed by no rain – resulting in drought – the next. Some of the negative effects of these changes, when considered alongside the life cycles of bees and other pollinators, are tangible.

A newly established bumblebee or solitary bee nest situated beneath the ground is not going to survive a flood; when rainfall is heavy and constant, bees cannot get out to forage, and if they manage to go out between downpours, it is extremely difficult for them to access the nectar and pollen in flowers filled with rainwater, so they are more likely to go hungry, and their nests to fail.

The lack of available pollen and nectar causes more problems for some species than others; honeybees, for instance, are able to feed on the honey and pollen stores they have built up in their hives for as long as these provisions last, and bumblebee colonies can survive for up to a few weeks on the nectar they have stored in their nests. Solitary bees, however, do not store honey, pollen, or nectar, so when food is scarce, they starve and die.

Dry conditions are no better. Increased temperatures, or extreme or prolonged droughts, cause flowers to wither and die, leaving little or nothing for insects to forage upon. But long before plants reach this stage, they begin to respond to the stresses imposed upon them, through lack of water and changing soil conditions, by producing less pollen, less nectar, and fewer flowers. This modification allows the plants to use what resources they can still draw upon to survive rather than to reproduce.

Yet flooding and drought are not the only challenges faced by pollinators because of climate change, and some of the problems are more complex and subtle than those I have just mentioned. As climate change brings with it ever-increasing extremes of weather, new patterns of behaviour are beginning to emerge amongst wildlife as it adapts to cope.

Insects are already responding to our warmer winters by changing their behaviour and life cycles. Some are shifting their geographical range to match new climatic ranges, as witnessed by once uncommon species, such as the Hummingbird Hawk-moth, becoming regular visitors in gardens both in Britain and Ireland.

Birds like little egrets and blackcaps are also migrating in greater numbers to our shores. However, according to recent scientific studies, it appears that some bumblebees in Europe and North America, whilst decreasing the southern extent of their ranges, are for some as yet unknown reason not expanding their range northwards. This means they are being squeezed into a narrower corridor.

The Great Yellow bumblebee, which is hanging on in the northernmost edges of Britain and Ireland, simply has nowhere else to go. There are winners and losers, but for many species, climate change is happening too quickly for them to adapt.

It is worth mentioning that some bees are able to switch into survival mode when they need to, instead of putting their energies into reproduction. Just as plants stop producing pollen and nectar when they are stressed, so do honeybee queens stop laying eggs when there is insufficient pollen available to feed their larvae. But what of the many other bee species, which only have a slim window of opportunity to provision their nests and lay their eggs? What will happen to them if the plants they rely on do not come into flower during the short time they are on the wing?

The timing of biological events such as insect emergence and plant flowering times is called *phenology*. Scientists are conducting research to find out how climate change is affecting pollinators, like my Hairy-footed Flower bee, at a species level.

It is becoming increasingly apparent that some flowering plants and the insects that pollinate them are beginning to fall out of sync. All rely on climatic signals and cues to one degree or another, but some use daily temperatures as cues – the onset of autumn frosts, for instance, is needed to trigger bud formation in some plants – whilst others use the length of daylight hours. If two species that depend on each other use different cues, or respond differently to the same cue, then both are in trouble.

Here in Britain and Ireland we do not need scientists studying phenology to tell us about the changes in the nesting behaviour of certain bumblebee species. Whilst it might still be unusual to find bluebells flowering in early March or late August, or Hairy-footed Flower bee females foraging on snowdrops in mid February, it is becoming increasingly normal to see active colonies of bumblebees in the south of England in the winter – a phenomenon that was, until quite recently, unheard of.

Instead of going into hibernation over the winter as they always have done in the past, some Buff-tailed bumblebee queens have for some years now been establishing new colonies in the autumn and producing workers that continue to forage throughout the winter months. And in the warmest parts of southern England, there has been evidence that these colonies produce reproductives during winter months, which means this species is becoming active all year round.

When I first started planting specifically for bees, the advice on offer was mostly about what to plant in spring, summer, and autumn, but to sustain these new populations of winter-flying bumblebees and other insects, we need to think about providing extra forage in the cold winter months, too. We can help them by planting winter-flowering plants such as mahonias, winter-flowering honeysuckle, heather, hellebores, snowdrops, and winter aconites. All these plants are winter hardy, so can cope well with frosts and light snowfall. Plant them, and they will provide life-saving nectar for winter-flying bees.

With the exception of Buff-tailed bumblebees and one or two others, the majority of bee species in Britain and Ireland still, mercifully, remain inactive during the cold winter months. But when spring arrives, bringing with it warmer temperatures and longer daylight hours, life begins to stir, and spring is coming earlier. Honeybee workers break away from their winter clusters, and the first of our queen bumblebees begin to emerge from hibernation or diapause, as bulbs and other plants begin to flower. If the temperatures are warm enough, this emergence could occur as early as February. Our earliest flying solitary bee species rarely appear much before March, but with weather fluctuating more wildly because of climate change, we may see more unusually warm weeks in January and February, too.

We have no idea yet how individual plants and pollinators will adjust and adapt to climate change, but the fact that some are already out of sync just compounds the myriad other challenges – from intensive farming and habitat loss, to pesticides, disease, and invasive species – that pollinators, and all the other creatures on this planet, already have to cope with. I hope and pray that governments and business leaders will wake up to this emergency and begin working together in earnest to reduce carbon emissions, which are, without doubt, the leading cause of the dramatic changes in Earth's climate.

In the meantime, we can all do our bit at home to help mitigate the effects of climate change and other threats to biodiversity, by reducing our own carbon footprint and waste, taking care of the soil, finding alternatives to pesticides, and planting, not just flowers but also meadows, shrubs, and trees. Planting, planting, planting, as though our lives depended on it. Because they do.

It is never too late to be the change.

Reflections

A s I come to write the last chapter in my book, I am drawn to revisit the Malvern Hills that I love and miss so much. Since I lived there for a few years as a child, circumstances have kept pulling me back, and I have felt a greater sense of belonging when I am on these ancient hills than I have in any other place. As a child, roaming the slopes and woods around our house, I felt safe and protected, and as an adult, being on the hills helped me find much-needed calm and balance during some very challenging times. It was here, too, over fifteen years ago, that I first understood how horribly disconnected I had become from the natural world, and I am grateful, every day, to whatever unseen energy it was that brought me that realisation.

Now I long to walk again the paths and tracks I used to tread, but with new eyes, ears, and awareness. I am curious to discover what I see and notice, that I might not have seen or noticed fifteen years ago. I am also looking forward to reacquainting myself with some of the sights and sounds I grew to know so well before I moved away and came to live in Shaftesbury.

It is nearly five years now since I left my hillside home in West Malvern, and though I have visited friends in the area on numerous occasions, I have for some reason chosen not to walk on the hills on those visits. I think perhaps I needed to wait until I had grown new roots, in Dorset, before stepping through the kissing gate at the top of Lamb Bank, and back onto the hills. But I am as content now as I have ever been, perhaps more so, and happily settled in my new place. I feel ready to revisit the hills without fear of being overwhelmed by nostalgia, or a yearning to move back to them.

I ring my good friend Louise to ask if she is up for a house guest, and whether she might like to accompany me on a walk on the hills. My walk would not be the same without Louise, whose company I used to enjoy on so many of my wanderings. We try, and fail, to meet at the end of the summer, and again in autumn, but life keeps getting in the way, so it is not till winter, when everything is sleeping, that we finally manage to take our walk.

And so, on an overcast day in early December, we walk from Louise's house in West Malvern, up Lamb Bank, through the kissing gate, and onto the hill. The muddy track leads us, almost immediately, past the rosebay willowherb (*Chamaenerion angustifolium*), which grows directly above the house I used to live in. I loved this flowering plant long before I ever knew its name, or that it was a favourite of bees and other pollinating insects. I think I loved it for its audacity: its brazen 'pinkness', whilst everything around it remained green, and the way it turned up year after year in my little patio garden, ever hopeful that I might turn a blind eye and allow it to grow. (I did once, but never again because it overstayed its welcome.) It didn't need my garden; it has a whole hill to grow on here.

It is known in North America by the striking name 'fireweed' because of how quick it is to colonise the ground after wildfires. But I prefer 'rosebay willowherb', which is so dreamy to say out aloud, that I am always on the lookout for excuses to mention it. "*Aaaaah*, rosebay willowherb," I say nonchalantly to Louise, drawing out each syllable as we walk past it.

It begins flowering in the month of June, on wasteland, hillsides, verges, riverbanks, and railway embankments across Britain and Ireland, and continues to bloom through July and well into August, its tall spires of magenta pink flowers opening in succession, as foxgloves do, from the bottom up. In fact, rosebay willowherb is often mistaken for foxglove from a distance. Its leaves, which resemble the leaves of willow trees, grow in spirals around the stem beneath the flowers, and in autumn they turn a golden scarlet red. I remember looking out of my back bedroom window once, as the setting sun cast the dying embers of its rays directly across the tops of the autumn willowherb, and thinking, for a moment, that the hill might be on fire.

I saw my first Elephant Hawk-moth on this patch of willowherb. So weird and exotic looking is this moth that I thought it may as well have come from another planet. It flew away before I had time to photograph it, which was most frustrating because I doubted at the time that anyone would ever believe, without documentary evidence, that I had encountered a giant pink moth. In Malvern. On the rosebay willowherb.

The willowherb has all gone over now, of course, but not before dispersing its feathery, spiderweb seeds – up to twenty thousand per plant – all over my old garden, and as far beyond as the wind can carry them.

We continue south along the well-worn track that circles this end of the hills. Here, in the dappled summer shade, I used to chase Speckled Wood butterflies. However carefully I approached them, they always flew up just as I got them perfectly framed in my viewfinder. I chase them still, but take their pictures now from afar, and crop them later on my laptop.

There is more vegetation on the hillside than I remember, even for this time of year. Below the path is mostly bramble, holly, and ivy, but above, where the hill rises steeply up to the ancient granite escarpment, bright young birch saplings stand proud and tall. And here, amongst all these bright young things, stands old man willow. He looks tired. We pause for a while with our backs to the willow and birch, and look back down across the village of West Malvern. The bluebell banks are blue no more, but covered in browning bracken. Beyond the bluebell bank is Whippets Brook, and a woodland ridge beneath which lie the villages of Cradley and Mathon. Looking farther into the distance are the hills of Herefordshire and the Black Mountains of Wales. This is a Narnian landscape if ever I saw one. C.S. Lewis went to school in Great Malvern, and surely would, at some time or another, have walked along this very same path and seen these very same hills and mountains. Sometimes, if I half close my eyes, I am quite sure I can see, amongst them, the castle of Cair Paravel.

Next we come to a stand of mature sycamore. They grow densely in the steep valley below the path, but more sparsely above – not because of the environment, but because the trees are thinned out by the Malvern Hills Trust. Growing on these trees we begin to see mosses and lichens, which are so much more conspicuous when the trees are bare than in the summer months. The ground is soggy and damp, covered in leaf litter and broken twigs, a haven for hibernating insects. As we pass through the sycamore, and into an area more exposed to the sun, the edges of the path to our left begin to bank up. The ground has become grassier – quite tussocky, in fact – providing perfect cover for small rodents' nests and runs. Bumblebees nest in these banks, too. I do not think there was a single year, once I started looking for them, that I did not find at least one or two bee nests in this area, mostly the homes of Buff-tailed and Red-tailed bumblebees. I am sure Garden bumblebees have nested somewhere nearby, too, but I have never found their nest.

As we climb higher I notice there is far more holly than I remember, but that may be because it is the only tree in leaf. Perhaps, like the moss and lichen, it is just more obvious in the winter months. There are very few berries on any of the larger established hollies, so I guess the redwings and fieldfares have already been through this way – unless these are male trees, which do not have berries. Louise and I try to recall if we have seen any of them flowering in the past. We agree that the largest tree, right beside the path, has had berries before. But there had been a heat wave over the summer, and very little rain. Chances are that this tree dropped its flowers during the dry spell to conserve water. Come to think of it, I haven't seen as many berries on the hollies by my home in Dorset this year as I usually do. I don't yet know my new patch as well as I do these hills, though I am eager to know it better.

We are walking on a level stretch of path now. It is more open here, with less shade, and along the edges of the path are low-growing plants: mouse-ear, ground ivy, herb Robert, dandelion, and nettle. I used to come and pick the young leaves from these nettles, and take them home to make tea and soup. For a couple of years running, a clump of hemp-nettle also grew in this spot, but then it disappeared and I have never found it again, either here or anywhere else.

The path becomes steeper, and as we climb higher we begin to see rowan (sometimes known as mountain ash) amongst the birch. Both rowan and

birch are *pioneer trees*, which means they are quick to plant roots in challenging environments, such as rocky ground and high altitudes, where
most woodland trees would struggle to survive. The rowan are quite bare
now, but earlier in the year they would have been covered in dense clusters
of creamy white flowers, and then laden with bright orange-red berries,
long since eaten by the birds. There are the beginnings of a few scratchy
patches of gorse farther up the hillside to our left, and some oak saplings,
still hanging on stubbornly to the last of their crinkly brown leaves. Near
them are a couple of spindly, old elders bound together with ivy. And more
holly, so much holly.

Beneath us the trees are almost entirely sycamore. Some of the more
mature trees stand alone, the ground beneath them inhospitable and
unwelcoming, whilst other, younger trees share their space with thickets
of bramble and thorn. In contrast, the slopes above us appear to have been
managed. The differences in the vegetation above and below the path are
really quite marked, but it occurs to me that I would probably not have
noticed these differences, or any of the individual trees, when I walked
this path on my way to work fifteen years ago. Would I even have seen the
differences between the elder, oak, sycamore, and rowan? Probably yes – if
you had lined them up and given me time to compare them. But separately,
apart from one another? I would have recognised the oak, with or without
its leaves, but the others I would have been able to separate and describe
only by their height and perhaps their girth. Now, if I came across a tree I
hadn't seen before, I may not be able to name it, but I would certainly be
able to describe its leaves, flowers, bark, and overall shape in far greater
detail than before. This is quite a revelation; my ability to observe and
describe nature has improved significantly. This must be thanks in part to
the online Botany course I started, but it is surely also due to my taking the
time now to notice, look, and listen.

The path curves sharply round to the left, past a large, healthy-looking
ash, before heading in a more southeasterly direction. The ground levels off
again as we reach one of my favourite stretches of the walk, the hawthorn
hedge. This ancient hedgerow forms a distinct boundary between the hill
and Joiners Meadow. When it blooms in May, it is heaven for all manner of
bees and other flying insects. According to the Woodland Trust, hawthorn
supports up to three hundred different species of invertebrates, as well as

providing shelter for numerous birds, making hedgerows such as this some of our most valuable if under-recognised homes for native wildlife.

Louise and I once celebrated Beltane – May Day – by dressing the whole length of this hedgerow with yellow and white ribbons. There are always big celebrations in Malvern on May Day. The hills and surrounding area are famous for their abundance of natural springs, and every year, around the bank holiday weekend, the town holds a Well Dressing Festival, during which more than fifty springs, wells, and spouts are decorated by community groups, schools, individuals, and families. I do so miss collecting water from these springs. It just doesn't taste the same from a tap.

If we were to walk a little farther on from the hawthorn hedge, following the same path and keeping the hill to our left, we would come to the spring at the top of Westminster Bank, where I used to collect most of my water. But we are not collecting water today, so instead we turn back on ourselves and cross over the track to the path that takes us up between Table and Sugarloaf Hills. Here we meet another walker, the first we have met since we began our walk. It is rare for the hills to be so empty.

At this stage we are so busy chatting and reminiscing that we almost walk straight past an old friend. Not a human friend, but another hawthorn, this time a single solitary tree, whose gnarled trunk and twisted branches have been shaped over the years, perhaps centuries, by the prevailing south-westerly winds. No other tree allows the wind to mould it quite like the hawthorn, and I am always struck by how dramatic this hawthorn looks as it somehow manages to hold its own in the exposed landscape.

There was a time, ten years or so ago, when this rather ancient-looking tree was not alone. Other trees grew on this slope, including some oak and an elder. None of them were 'ancient', but some of the oaks were at least a hundred years old. They were felled to create more habitat for the declining population of ground-nesting birds. To this day I cannot reconcile, in my head, how you decide which is more important, birds or trees. I remember protesting vehemently at the time against the felling, and I would do so again today. Yet, if there is one thing I have learned over the last decade, it is that decisions about wildlife conservation and management are never as simple, or as cut and dried, as they appear, or as you want them, to be.

We say our goodbyes to the hawthorn, and carry on up the path until we reach a small quarry. The hills are full of old quarries, including the one from which I used to collect frog spawn and newts when I was a child. Many have filled with water; this one has not. It is actually more of a dugout than a quarry, and provides a nice sheltered spot for a weary walker.

It was here, some five or six years ago, that Louise and I made our first nature mandala together, using fallen leaves, berries, conkers, acorns, crab apples, and other finds we had gathered on our walk that morning. A mandala is a Hindu and Buddhist symbol representing the universe, circular in design. You work from the inside outward, maintaining symmetry and balance, adding whatever you want until you are happy with what you have created (or have used up all your materials). So long as it is circular, there is no right or wrong way to make such a thing. Some people paint or draw them; I like to make them with nature finds.

We were so proud of our mandala! It took us hours to complete, and was, we thought, the most beautiful piece of nature art either of us had ever

seen. Since then, I have experienced great joy making mandalas at different times of the year, in different locations, and with different natural materials. Afterwards, I leave them for other walkers to find – and eventually, for the wind to blow away. I made some last year, along Pine Walk in Shaftesbury, four or five small ones, each different. A couple contained berries from our allotment, and I remember enjoying the thought that they were more likely to be eaten by passing creatures than to be blown away by the wind. For me, this is a beautiful way to revel in nature's bounties, mark the changing of seasons, and reflect on the passing of time.

There are other ways I have marked the changing of seasons on the hills. When I first moved back to Malvern, my son, James, came a few times to celebrate the winter solstice with me. The first winter he joined me, the wind was so fierce that we had to crawl our way up the side of North Hill to catch the sunrise. At one point, I was worried we would be blown off the hill if we dared to stand up. It was worth it though; it always is.

The following year when James came, there was a heavy snow, and ours were the very first tracks – human or otherwise – in the drifts that morning. Each time we put our feet forward, they sank so deep into the snow that we had to reach down and grasp the sides of our wellies to pull them out again. It was very slow going, and we didn't actually make it to the top in time to see the sun rise, but the journey was a lot of fun.

I saw in the millennium here, too, with my family. We walked up to the top of the Worcestershire Beacon to have a bird's-eye view of the firework display at the Three Counties Showground. It was bitterly cold and extremely icy, so we mostly slipped our way up the hill, barely getting there in time for midnight. What we hadn't bargained on was a thick heavy cloud enveloping the hilltop. We couldn't see a thing – not the fireworks, the moon, the stars, or each other – but it was a great place to see in the new century.

These hills are truly ancient, by far the oldest in England and Wales. The mainly granite rock from which they were shaped was formed some 600 million to 1,000 million years ago, in the Precambrian aeon. Their energy is tangible. I felt it as a child, and I feel it again now. I bet they could tell you some stories, these hills.

After walking up from the quarry, Louise and I stop to rest awhile on the rustic wooden bench that sits atop the so-called Saddle. This is a meeting place between paths, tracks, and hills, where walkers stop and check their maps and families pause for picnics. To the east lies the town of Great Malvern, and beyond that, the city of Worcester and the Vale of Evesham; to the west, Herefordshire and Wales; to the north, North Hill; and to the south, its summit the highest of all the hills in this range, the Worcestershire Beacon. I have always preferred the more wooded areas on the lower slopes of the hills, so don't often walk higher than this area, but this flat, in-between area holds special memories for me.

You would not know it at this time of year, whilst the ground is cold and hard, and nothing on the hill is flowering other than the odd gorse, but bees nest in this spot. If I were to come back next year in March or April, I know I would find good numbers of one of Britain and Ireland's earliest-flying solitary bees, the Large Sallow Mining bee (*Andrena apicata*). Also lurking by the entrances to the mining bees' nests would be its cuckoo, the Early Nomad bee (*Nomada leucophthalma*).

For years I must have walked straight past these early spring bees, my mind focused on more – or as it turned out, less – important things. But then one day I noticed them, an aggregation of solitary mining bees, *A. apicata* (though I didn't know that was what they were; they were identified as such by Ian Beavis from the photographs I posted on Twitter). After filling my Twitter feed with photos of the bees, one of my entomologist contacts suggested I should also look out for the cuckoo bee that lays her eggs in this particular mining bee's nest.

I headed straight back up the hill, and sure enough, there she was, with her shiny, almost hairless, yellow-and-black striped abdomen, marked here and there with distinct patches of red – my very first solitary cuckoo bee. I was delighted. Every day for weeks, I walked up to see those bees, lying on the ground to get as close to them as I could without blocking the entrance to any of the nests. I took my life into my own hands doing this, because it turned out this was a very popular track amongst mountain bikers.

I marvelled at the time that the bees were able to survive such traffic and wondered why on Earth they had chosen this location. But *A. apicata* loves nothing more than to nest in areas of flat, compacted soil in exposed sunny positions such as this one. They also collect pollen almost exclusively from pussy willow flowers. And guess what grows on the other side of the wooden bench, just below the Saddle, at the top of the path that leads down the hill to Great Malvern? Pussy willow.

This spot is even more special for another reason, a reason that used to bring me back here daily throughout the hot summer months, long after the mining bees had stopped flying. This is skylark territory.

I do not remember when I first noticed the skylarks singing on the Malvern Hills, but once I did, their song captivated and consumed me like no other sound, filling me with not only a great sense of hope and joy, but a tremendous sense of freedom as well. How such a tiny creature can fill the whole sky and the length and breadth of hills with its song is beyond my understanding. I cannot think of anything sweeter, or more pure, than the song of the skylark as he flies, suddenly and unexpectedly, up from the ground, ascending to the very heavens themselves, where he hovers on the wind, so high that he is scarcely visible, but still singing, all the time, singing. More than any other living thing, the skylark embodies the spirit of the hills for me.

Much as watching bees led me, gradually, to start noticing the other insect visitors to our garden and allotment, it was listening to skylarks on the Malvern Hills that was my first introduction to the world of nature's soundscapes. The Malverns, with their mosaic habitat of bracken, scrub, grassland, bare rock, and trees, are a wonderland of different sounds. There are, of course, the skylarks, but also other birds, each with its own distinctive song or call. I learned to recognise the calls of the green woodpecker in the open, grassy areas, and that of chiffchaff, other warblers, and the tawny owl in the woodland. And it was the buzzard's mew I could make out as the birds circled high above the hills. Closer to home, I became familiar with the resident robin and blackbird, so much so, that whilst I was at my most tuned in, I could tell by its song if a neighbouring robin or blackbird had wandered onto their territory.

Today, though, I do not hear any skylarks. I would not expect to at this time of year, but sadly I hear later from friends that skylarks no longer nest on these hills. They have gone.

As Louise and I descend from the Saddle and head for St Ann's Well, we are surrounded again by silver birch, growing densely in the valley beneath us as well as the slopes above. Their feathery branches and twigs are purple against the grey-blue sky. Here, in the summer months, I used to be drawn off the path and onto the wooded slopes, following what I thought was the hum of honeybee swarms. Year after year I was surprised to find it was the hum of hoverflies. I never did work out what they were doing here, or why their collective buzz was so loud that I could hear it from the path. Perhaps they were feeding on the sap of the birch or, more likely, on the aphids that feed on the sap.

By the time we reach the bottom of the path, my knees are beginning to complain, so when we get to the knoll above St Ann's Well, we decide to forgo our trek into town for a cup of tea and instead walk back up Happy Valley and over North Hill to Louise's house on the western side of the hills.

The path through Happy Valley has a completely different feeling to any other path on the hills. Edged with trees, it is more like an avenue, or a ride, and straight away I hear the familiar sound of the little brook that runs down the left side of the ride. I imagine it must originate from a spring in the side of the hill, but I do not know where. At the top of the valley we pass the pussy willow and an old stone footpath marker where I once found a dug-up bumblebee nest.

We rest again, after we have climbed the steep path that leads from Happy Valley to North Hill, and take a moment to look back towards the Worcestershire Beacon. A group of walkers are approaching the summit, but these are the first we have seen on this side of the hills. Beneath them, on the eastern slopes, I can pick out the extended patches of bilberries that expand year upon year, covering ever more ground. I used to spend hours sitting amongst those bilberries in the hope that I might see Bilberry bumblebees. I never did. It was a long shot, as, to my knowledge, this species has never been recorded on the Malvern Hills. Clearly, you need more than bilberries to sustain a Bilberry bumblebee population.

However, one autumn, in the wooded area just beneath the bilberries, I watched an enormous Buff-tailed bumblebee queen digging a tunnel in the soil around the roots of one of the trees. This was before I knew anything about the biology or behaviour of bumblebees, so I had it in my head that she might be digging her nest. I was transfixed, thinking of the eggs she would lay there. It was only after I got home that I learned she was digging herself a safe place to hibernate over the winter.

I wonder how many nests of hibernating bumblebee queens Louise and I have walked over today. Not only bumblebees, but other creatures, too, will be slumbering in their hibernacula, amongst the roots of trees and plants, beneath leaves and rocks, or in other nooks, crannies, and crevices, safe from the winter frosts until they are touched by the warmth of the sun, and the longer daylight hours cause them to waken from their long winter naps.

Plants also lie dormant beneath us, waiting in their dark invisible underground beds for the return of the light – bluebells, bracken, wild garlic, and more. Spring blossoms seem a long way away, but we are approaching the winter solstice, the shortest day, so the days are soon to lengthen. Still, the dark winter months are important in their own right, as they give us all time to rest and rejuvenate, to hunker down; for us humans, to drink soup and hot chocolate, and reflect on the year that has just gone by. This is the time to make plans and resolutions for the coming year. Spring, and new life, will be here soon enough.

I feel exhilarated as we reach the top of North Hill, and blessed that the weather has been so kind to us today. Though it had been threatening rain when we set off, the sky above is now quite clear, and there is a kestrel hovering right above us. This is a good hunting ground for kestrels.

The views from North Hill are breathtaking. We used to walk from our home on the other side of the Worcestershire Beacon to this hill on Sundays after lunch when I was child, my father, my three younger brothers, and myself. And when we reached the top we used to slide all the way back down again on our bottoms; not on the path, but on the long slippery grass that lends itself so well to this purpose. We used to roly-poly, too, but I didn't enjoy that so much because it made me dizzy and I never knew where exactly I would end up.

I am sure there used to be more gorse growing up here. There were harebells, too, in the summer, growing alongside the paths on this part of the hills. It always surprises me to see something as delicate and fragile as a harebell thriving in areas as exposed as this.

Louise beckons me with her arm to join her where she is standing. Beneath us, along the ridge that runs from Eastnor to Coddington, something quite unexpected and magical is afoot. A thick, white ribbon of mist is snaking its way slowly down towards the wooded ridge. Like a silken serpent, it drapes and envelops the entire ridge, leaving only Bradlow Knoll, the highest point of Frith Wood, exposed, before stretching away from the ridge and winding its way down into the Colwall valley, where it forms a pool of cloud before dispersing, wisp by wisp, into nothingness. In all the years I have walked on these hills, I do not think I have seen such a thing. Surely I would have noticed it if I had? Either way, I am glad to have witnessed it today. The phenomenon, I discover later, is sometimes called 'dragon's breath'.

We had planned to finish our walk on the steep slope that leads down from End Hill and joins the path from which we set off, but my knees are really struggling now, so we retrace our steps, taking the gentler route back. I regret that I will now miss out on visiting my favourite tree, the beautiful silver birch whose lower bough I used to sit upon when I needed a hug. Silver birch is often referred to as 'the Queen of the Woods,' but this tree is Queen of the Hills. Unfortunately, though, she grows halfway down the steep slope we wish to avoid. In hindsight, we should have walked up that way. However, I will return.

I have so enjoyed walking on the hills again. It is hard, sometimes, to go back to a place you have loved so much, for fear that it might make you feel confused about whatever place you now call home. But I do not feel these things today. I feel no dissatisfaction, no sadness; instead I feel liberated. I feel like the skylark, as he sings his heart out from the heavens above the hills; the wind, rushing through the leaves in the trees; the stream that cascades over the rocks beneath the cabin I stayed in at Ashbury; the gannets as they plunge like torpedoes into the sea; and the bees that buzz and hum in our garden and on our allotment.

This tremendous and unexpected sense of freedom comes from knowing that the oneness I experienced for the first time when I was sitting with a dandelion in my friend Rachel's garden in Stroud, can be experienced anytime and anywhere. It matters not whether you are in a familiar landscape, or in a faraway place, whether you are surrounded by trees, mountains, rivers, or the vegetables on your own allotment. Having a relationship with the rest of nature is about opening our hearts, our minds, and ourselves, knowing that we can, if we wish, rekindle our lost connections, because somewhere deep inside us all, there lives a little spark of 'wild' just waiting to be ignited, or re-ignited. All it needs is for us to allow it room to grow.

I have certainly not found, nor do I seek to find, some kind of nirvana. I am firmly rooted in a human body, with a sometimes overactive human mind, and crazy human emotions that often get the better of me. And for a 'nature lover', I am deeply flawed: I prefer sleeping in a comfortable bed to sleeping on a forest floor; I do not like putting my head beneath the waves when I am swimming in the sea; and though I love bees more than I ever thought it possible to love a flying insect, I am still a little afraid of being stung by one.

But since rediscovering the awe and wonder I felt for the natural world when I was a child, I have been blessed with the ability to see miracles in everything around me, in the big things, the little things, and the ever-so-slightly scary things. Every single day, no matter what horribleness is happening in my world, or the wider world, I feel blessed in the knowledge that I can find solace, refuge, strength, and joy in an instant, just by stepping outside.

I still love reading books about nature, and I enjoy attending courses run by people who can teach me about birdsong or bee behaviour, or how to recognise a tree by its winter twigs and buds. These ways of learning suit me well, and it makes no sense to discard them. But I also love learning about wildlife from the wildlife itself. I have stopped trying to understand why I spent so many years not attending to nature, because it really doesn't matter. I don't need to know. I am just happy to have found joy again in my surroundings, as I did before I somehow got lost for a while in the merry-go-round of life.

Just one thing bothers me, and that is the detrimental effect I have had in my lifetime on the wild creatures I have grown to love and cherish. I cannot turn back the clock, but I am aware that, alongside all my efforts to plant and create habitat for bees and other wildlife, I need also to increase my efforts to reduce the negative impact that my everyday life has upon those same creatures. We all need to do so, before it is too late.

Loving nature means so much more than enjoying and appreciating it, and more than photographing and writing about it. It means standing up for it, fighting for it, and accepting it, unconditionally, warts and all. We ourselves *are* nature, but by treating the rest of nature the way we do, as if it is 'separate' to us, we endanger the wild animals, plants, and landscapes which we act so carelessly towards, as well as ourselves. It is only when we realise that we are a part of nature, rather than apart from it, and behave accordingly, that real change is likely to happen.

There are many wonderful moves afoot, and projects under way to 'rewild' the land: to allow forests to regrow, and to reintroduce creatures such as beaver, lynx, and wolf to habitats they once roamed freely. I applaud these movements and hope with all my heart that they succeed. But I would also like to see a shift in attitudes towards the wild creatures already in our midst, most especially the smallest amongst them. I cannot see us accepting the presence of large predators whilst we have yet to accept the presence of the invertebrates, reptiles, and small mammals that so many of us still fear. And spare a thought, if you will, for the 'weeds', those stalwarts such as ragwort, creeping thistle, and dandelion that provide such invaluable forage for all manner of insects, many of which rely on these plants for their very existence. Too much of wildlife has been deemed 'pests' or 'thugs'. Far from recognising their value to pollinators and other creatures, and allowing them space to grow, we wage war upon them and their kind.

Finally, whilst all this is going on, perhaps we might also take the time to look inwards and begin the process of rewilding *ourselves*, by spending more time with nature – without an agenda, and without attaching ourselves to the outcome. We might then pay attention to the feelings and emotions this brings up within us, noting and accepting anything new. Do we feel relaxed or tense? Fearful or 'at one' with our surroundings? And how do these responses change over time and with ongoing practice? If we do not come to know and embrace the wildness *within* ourselves, I do not know in all truth how we can fully embrace the wildness *without*. We need only take a few, small steps to begin the process of rewilding, dipping our toes in the shallow end of the wonderful pond of life, so to speak. But as my mother used to tell me, 'From small acorns, great oaks grow.' And she was right.

A honeybee colony, sixty thousand strong, is capable of pollinating an entire apple orchard. But a solitary Red Mason bee, working all alone, can pollinate as many flowers in that apple orchard as over one hundred honeybees. We all have our part to play, individually *and* together.

Humans are truly remarkable beings. Together we have the power to effect great change. But we can make a difference as individuals, too. So whether you plant a single lavender plant on the balcony of your flat, a wild flower meadow in your lawn, a patch of phacelia on your allotment, or a field full of flowering trees on your hundred-acre plot, you will be helping the beautiful pollinators without which our lives would be so much poorer.

And next time you see a bee, don't forget to thank it.

ACKNOWLEDGEMENTS

The poet John Clare 'found his poems in the fields'. I am not a poet, but I, too, found inspiration in the fields, as well as in the hedgerows, woodlands, valleys, and vales, and, of course, on my beloved Malvern Hills. Can you 'thank' a range of hills? If you can, I do, together with the bees, trees, birds, wild flowers, and all the other wondrous creatures which have charmed their way into my heart and into this book. It is for them, and because of them, that I am moved to write.

It has taken me nearly ten years to complete this book, and during that time I have learned as much through the writing as I have through my guides, the bees. It has been a steep, uphill learning curve, more challenging at times than running a marathon, but the sense of achievement has made every single sleepless night worthwhile.

One of the biggest challenges has been writing about specific aspects of bee biology or behaviour that I have not witnessed myself. To this end, I enlisted the help of those far more knowledgeable than me – entomologists and other scientists – who kindly agreed to check the more technical pieces in my book for errors. To these experts – Richard Comont, science manager with the Bumblebee Conservation Trust; Steven Falk, author of *Field Guide to the Bees of Great Britain and Ireland*; Jeff Ollerton, professor of biology at Northampton University; Stuart Roberts, former chairman of the Bees, Wasps and Ants Recording Society; John Walters, consultant entomologist and naturalist; and Matt Shardlow, CEO of the charity Buglife – I owe an enormous debt of gratitude. Thank you so much, not only for your tremendous help with this book, but also for correcting me over the years when I got things wrong (which I have, often); for the generosity you display in sharing your knowledge via social media; and for all you do to help demystify the science for us everyday folk. You are, each and every one of you, amazing, and I salute you.

Others have advised me, too. I owe a huge debt of thanks to Phil Chandler, author of *The Barefoot Beekeeper* and fellow 'Friend of the Bees', for helping me understand more about honeybees (and, more important, for organising the event where I met my husband). Thank you also to others who spoke with me, or communicated via email, about their important work on, and on behalf of, insects, plants, wildlife and wild places: Kate Bradbury, Jim Cane, Sue Clifford, Rachel Corby, Dave Goulson, Angela King, Ron Rock, Matt Somerville, Isabella Tree, and John Walters.

Any mistakes and inaccuracies that remain are entirely my own, and for these, I humbly apologise.

There are also those who helped and advised me before my book even began to take shape: Alys Fowler and Will Rolls, thank you for the benefit of your own writing experiences, and helping me weigh up the pros and cons of publishing versus self-publishing; Shaun Chamberlin, for reading my early scribblings, and enjoying them enough to introduce me to Chelsea Green Publishing; Peta Nightingale, for helping me negotiate my way through the world of contracts, and for the best advice ever, that is, to 'read everything out aloud' before submitting it; Judy Napper, for proofreading my book proposal; and Charlotte, my daughter, for making my proposal look so professional.

I am deeply indebted to Margo Baldwin, president and publisher of Chelsea Green Publishing, for taking a chance on me and publishing this book. Heartfelt thanks, as well, to UK managing director Matt Haslum, editor Michael Metivier, production director Pati Stone, and the rest of the team at Chelsea Green. I have really enjoyed working with you. Also, thanks to Eliani Torres for your copyediting skills and suggestions.

Writing the odd blog post is one thing, but writing a book is another thing altogether, and this I would not have achieved without the help of Robin Dennis, my developmental editor. Robin is something of an alchemist. Like honeybees take nectar and turn it into honey, she took my chaotic ramblings and turned them into a book. In fact, in another life, I think she might have been a honeybee. Thank you, Robin, for upping my game, for steering me through processes that took me way out of my comfort zone, and for your patience and forbearance.

I have known from day one that I wanted wildlife artist John Walters to illustrate my book, and I was over the moon when he agreed. I could not be more delighted with how his paintings bring each creature to life. Thank you, John. Let's do this again!

For inspiring and delighting me before, during, and after I put my own pen to paper, I thank the nature writers whose books I enjoy so much. In no particular order: Hugh Warwick, Rob Cowan, Robert Macfarlane, John Lewis-Stemple, Miriam Darlington, Mark Cocker, Peter Marren, Amy Liptrot, John Lister-Kaye, Kate Bradbury (again), Simon Barnes, Jon Dunn, Tim Dee, Mark Avery, Tom Cox, Nicola Chester, Will Cohu, Dave Goulson (again), Michael McCarthy, Amy-Jane Beer, Matthew Oates, Patrick Barkham, Mary Colwell, and more. Also, poets and writers from days gone by: John Clare, Roger Deakin, Jean-Henri Fabre, and Gerald Durrell, to mention but a few.

'No man is an island' said John Donne, and never were truer words spoken. Without the help and encouragement of my friends and family, I would have long ago abandoned any idea of writing this book. So thank you to all my amazing friends, old and new – you know who you are. I love you and feel blessed to have you in my life. Special thanks to Natalie, Louise, Rob and Zoe, and Jon and Tanya, for always being there and supporting me through thick and thin. And to my Facebook friends and Twitter followers: Thank you, too, for accompanying and supporting me on my journey, and for filling my world with such beautiful, uplifting posts and wildlife photographs. A special mention goes to Ian Beavis, for all your help and guidance over the years. Sue Clifford and Angela King (again): I cannot thank you enough for your advice and encouragement, and for all the coffee and croissants; you inspire me to keep fighting for what I believe in.

My wonderful children, James and Charlotte, and my daughter-in-law, Holly, thank you for believing in me; for seeing me through some seriously challenging times; for unfailingly supporting and encouraging me; and for patiently putting up with years and years of my talking incessantly about bees. I love you. Thank you also, Charlotte, for teaching me to believe in myself again. Rob's daughters, Amber and Jade, thank you for being so excited and enthusiastic about this book. My grandchildren, Indy, Pippin, and Arrietty: Thank you for the light and joy you bring me. I love you to the moon and back!

Last, but far from least, my husband and best friend, Rob. Thank you for everything – for sharing the ups, and helping me through the downs;

for listening as I read these chapters out aloud, again and again; for making sure I eat; and for keeping all the other home fires burning whilst I was completely absorbed in writing. You are my rock. I love you more than words can say.

LIST OF ILLUSTRATIONS

Page 203   Orange-tailed Mining bee (*A. haemorrhoa*) swarming
           blackthorn (*Prunus spinosa*).
Page 215   Ivy bees (*C. hederae*) patrolling a cliff nest.
Page 219   Harlequin ladybird (*Harmonia axyridis*).
Page 229   Hairy-footed Flower bee (*Anthophora plumipes*) on
           snowdrops (*Galanthus*).
Page 239   Buff-tailed bumblebee (*B. terrestris*) visiting sallow
           (*S. caprea*) catkins.
Page 245   Nature mandala, Malvern Hills, Autumn 2011.

SELECTED BIBLIOGRAPHY

BOOKS

Alford, D.V. *The Life of the Bumblebee.* Hebden Bridge: Northern Bee Books, 2009.

Benton, Ted. *Bumblebees.* London: Collins, 2006.

———. *Solitary Bees.* Exeter: Pelagic Publishing, 2017.

Chandler, Philip. *The Barefoot Beekeeper: A Simple, Sustainable Approach to Small-Scale Beekeeping Using Top Bar Hives.* Lulu, 2015.

———. *Learning from Bees: A Philosophy of Natural Beekeeping.* Micro Publishing Media, 2012.

Comont, Richard. *RSPB Spotlight Bumblebees.* London: Bloomsbury Natural History, 2017.

———. *RSPB Spotlight Ladybirds.* London: Bloomsbury Natural History, 2019.

Corby, Rachel. *The Medicine Garden: Gather and Make Your Own.* Preston: Good Life, 2009.

———. *ReWild Yourself: Becoming Nature.* Stroud: Amanita Forrest Press, 2015.

Edwards, Mike, and Martin Jenner. *Field Guide to the Bumblebees of Great Britain & Ireland,* 6th ed. Lewes: Ocelli, 2018.

Fabre, J. Henri. *Bramble-Bees and Others.* Translated by Alexander Teixeira de Mattos. Whitefish, MT: Kessinger Legacy Reprints, 2010.

Falk, Steven. *Field Guide to the Bees of Great Britain and Ireland.* London: Bloomsbury, 2015.

Goulson, Dave. *Bumblebees: Behaviour, Ecology and Conservation.* Oxford: Oxford Biology, 2009.

———. *A Buzz in the Meadow: The Natural History of a French Farm.* London: Picador, 2016.

———. *A Sting in the Tale: My Adventures with Bumblebees.* London: Jonathan Cape, 2013.

Hanson, Thor. *Buzz: The Nature and Necessity of Bees.* London: Icon, 2018.

Kirk, W.D.J., and F.N. Howes. *Plants for Bees: A Guide to the Plants That Benefit the Bees of the British Isles.* Bristol: International Bee Research Association, 2012.

Maeterlinck, Maurice. *The Life of the Bee.* Translated by Alfred Sutro. Sydney: Allen & Unwin, 1916.

Martin, W. Keble. *The Concise British Flora in Colour*, 2nd ed. London: Ebury and Michael Joseph, 1976.

McAlister, Erica. *The Secret Lives of Flies*. Richmond Hill, ON: Firefly Books, 2017.

Michener, Charles D. *The Social Behavior of the Bees: A Comparative Study*. Cambridge, MA: Belknap Press, 1974.

O'Toole, Christopher, and Anthony Raw. *Bees of the World*. Richmond Hill, ON: Firefly Books, 2013.

Prŷs-Jones, Oliver E., and Sarah A. Corbet. *Bumblebees*, 3rd ed. Exeter: Pelagic Publishing, 2011.

Rose, Francis, and Clare O'Reilly. *The Wild Flower Key: How to Identify Wild Flowers, Trees and Shrubs in Britain and Ireland*, 2nd ed. London: Warne, 2006.

Sladen, F.W.L. *The Humble-Bee: Its Life History and How to Domesticate It*. Hereford: Logaston Press, 1989.

Stubbs, Alan E., and Steven J. Falk. *British Hoverflies: An Illustrated Identification Guide*, 2nd ed. Wokingham: British Entomological & Natural History Society, 2009.

Tautz, Jürgen. *The Buzz about Bees: Biology of a Superorganism*. Translated by David C. Sandeman. Berlin: Springer, 2009.

Tree, Isabella. *Wilding: The Return of Nature to a British Farm*. London: Picador, 2018.

White, Gilbert. *The Natural History of Selborne*. Oxford: Oxford University Press, 2013.

Wilson, Joseph S., and Olivia Messinger Carril. *The Bees in Your Backyard: A Guide to North America's Bees*. Princeton, NJ: Princeton University Press, 2015.

ARTICLES

Bischoff, I., E. Eckelt, and M. Kuhlmann. 'On the Biology of the Ivy-Bee *Colletes hederae* Schmidt & Westrich 1993 (Hymenoptera, Apidae)'. *Bonner Zoologische Beiträge* 53 (2005). 27–35. Cited in '*Colletes hederae* Schmidt & Westrich, 1993', Bees, Wasps and Ants Recording Society, http://www.bwars.com/bee/colletidae/colletes-hederae.

Cane, James H., George C. Eickwort, F. Robert Wesley, and Joan Spielholz. 'Foraging, Grooming and Mate-seeking Behaviors of *Macropis nuda* (Hymenoptera, Melittidae) and Use of *Lysimachia ciliata* (Primulaceae) Oils in Larval Provisions and Cell Linings'. *American Midland Naturalist* 110:2 (October 1983). 257–64. https://www.jstor.org/stable/2425267.

Center for Food Safety. 'Hidden Costs of Toxic Seed Coatings: Insecticide Use on the Rise'. Fact sheet (June 2015). https://www.centerforfoodsafety.org/files/neonic-factsheet_75083.pdf.

Goulson, Dave, Elizabeth L. Sangster, and Jill C. Young. 'Evidence for Hilltopping in Bumblebees?' *Ecological Entomology* 36:5 (August 2011). 560–63. https://doi.org/10.1111/j.1365-2311.2011.01297.x.

Ray, C. Claiborne. 'Tree Power'. *New York Times*, 3 December 2012. https://www.nytimes.com/2012/12/04/science/how-many-pounds-of-carbon-dioxide-does-our-forest-absorb.html.

Roy, Eleanor Ainge. 'Nigel the Lonely Gannet Dies as He Lived, Surrounded by Concrete Birds'. *Guardian*, 1 February 2018. https://www.theguardian.com/world/2018/feb/02/nigel-lonely-new-zealand-gannet-dies-concrete-replica-birds.

Strom, Stephanie. 'A Bee Mogul Confronts the Crisis in His Field'. *New York Times*, 16 February 2017. https://www.nytimes.com/2017/02/16/business/a-bee-mogul-confronts-the-crisis-in-his-field.html.

Williams, Paul H., and Juliet L. Osborne. 'Bumblebee Vulnerability and Conservation World-wide'. *Apidologie* 40:3 (May 2009). 367–87. https://doi.org/10.1051/apido/2009025.

ORGANISATIONS AND WEBSITES

AgriLand. https://www.agriland.co.uk.

Ancient Yew Group. https://www.ancient-yew.org.

Barefoot Beekeeper. http://www.biobees.com.

Bee Happy Plants. https://beehappyplants.co.uk.

Bee Kind Hives. https://beekindhives.uk.

Bees for Development. http://www.beesfordevelopment.org.

Bees, Wasps and Ants Recording Society (BWARS). http://www.bwars.com.

Botanical Society of Britain & Ireland (BSBI). https://bsbi.org.

British Trust for Ornithology (BTO). https://www.bto.org.

Buglife. https://www.buglife.org.uk.

Bumblebee Conservation Trust (BBCT). https://www.bumblebeeconservation.org.

Bumblebee.org. http://www.bumblebee.org.

Butterfly Conservation. https://butterfly-conservation.org.

Buzz about Bees. https://www.buzzaboutbees.net.

Common Ground. https://www.commonground.org.uk.

Richard Comont. http://www.richardcomont.com.

Rachel Corby, Gateways to Eden. http://gatewaystoeden.com.

Devon Wildlife Trust. https://www.devonwildlifetrust.org.

Dorset Wildlife Trust. https://www.dorsetwildlifetrust.org.uk.

Friends of the Bees. https://www.friendsofthebees.org.

Green & Away. https://www.greenandaway.org.

Honey Bee Suite. https://honeybeesuite.com.

iRecord. https://www.brc.ac.uk/irecord.

iSpot. https://www.ispotnature.org.

Knepp Castle Estate. https://knepps.co.uk.

Malvern Hills Trust. http://www.malvernhills.org.uk.

Natural Beekeeping Trust. https://www.naturalbeekeepingtrust.org.

NHS Forest. https://nhsforest.org/evidence-benefits.

Nurturing Nature. https://nurturing-nature.co.uk.

Jeff Ollerton's Biodiversity Blog. https://jeffollerton.wordpress.com.

Plantlife. https://www.plantlife.org.uk.

The Pollinator Garden. http://www.foxleas.com.

River of Flowers. http://www.riverofflowers.org.

Rosybee Plants for Bees. http://www.rosybee.com.

The Royal Society for the Protection of Birds (RSPB). https://www.rspb.org.uk.

Shropshire Wildlife Trust. https://www.shropshirewildlifetrust.org.uk.

The Species Recovery Trust. http://www.speciesrecoverytrust.org.uk.

Trees for Life. https://treesforlife.org.uk.

UK Ladybird Survey. http://www.coleoptera.org.uk/coccinellidae/home.

John Walters. http://johnwalters.co.uk.

The Wildlife Trusts. https://www.wildlifetrusts.org.

The Woodland Trust. https://www.woodlandtrust.org.uk.

Xerces Society for Invertebrate Conservation. https://xerces.org.

INDEX

Note: Page numbers in *italics* refer to illustrations; page numbers followed by *t* refer to tables.

INDEX

birch trees, 242–43, 248–49
bird boxes, Tree bumblebee nests in, 117
birds, neonicotinoid poisoning of, 100
bird watching
    Abernethy Forest, 136
    author's mother, 83–88
    Big Garden Birdwatch, 185
    birdsong identification, 91, 93, 97
    Howard, Rob, 95
    Isles of Lewis and Harris visit, 146
Bishnoi village, origin of tree-huggers, 175
Blandford fly (*Simulium posticatum*), 122
Blotch-winged hoverfly (*Leucozona
    lucorum*), 128
bluebells, 88, 89
blue tits, 116
bog asphodel, 143–44
Bombus distinguendus. See Great Yellow
    bumblebee (*Bombus distinguendus*)
Bombus hortorum. See Garden bumblebee
    (*Bombus hortorum*)
Bombus hypnorum. See Tree bumblebee
    (*Bombus hypnorum*)
Bombus jonellus. See Heath bumblebee
    (*Bombus jonellus*)
Bombus lapidarius. See Red-tailed
    bumblebee (*Bombus lapidarius*)
Bombus lucorum. See White-tailed
    bumblebee (*Bombus lucorum*)
Bombus monticola (Bilberry bumblebee),
    68, 132
Bombus muscorum. See Moss Carder
    bumblebee (*Bombus muscorum*)
Bombus pascuorum. See Common Carder
    bee (*Bombus pascuorum*)
Bombus pratorum. See Early bumblebee
    (*Bombus pratorum*)
Bombus terrestris. See Buff-tailed
    bumblebee (*Bombus terrestris*)
Bombylius major (Large bee-fly), 129
Botanical Society of Britain & Ireland
    (BSBI), 185, 203
Bovey Heathfield Nature Reserve, 160–68
Box-headed Blood bee, 37

boxing hares, 132
Bramble-Bees and Others (Fabre), 199
*British Hoverflies: An Illustrated Identification
    Guide* (Stubbs and Falk), 126
broad beans, nectar robbing from, 78,
    79–80
Broad-bordered Bee Hawk-moth (*Hemaris
    fuciformis*), 129–130
Brodie, Laura, 69
brood parasites (cleptoparasites),
    42–43, 104
    See also cuckoo bees
BSBI (Botanical Society of Britain &
    Ireland), 185, 203
Buff-tailed bumblebee (*Bombus
    terrestris*), 239
    allotment colony, 31–32
    attraction to lamb's ears, 194
    climate change-related shifts in nesting
        behaviour, 236–37
    comfrey visits, 75, 76
    courtship and mating behaviour, 68
    cuckoo bees of, 107
    emergence from hibernation, 19–20, 24
    nectar robbing, 76, 77
    nest in author's garden, 28–30
    patio garden visits, 125
    siting of nests, 24–25
    worker bees, 27
Buglife charity, 31, 162, 210
Bumblebee Conservation Trust (BBCT),
    30–31, 42, 151
Bumblebee hoverfly (*Volucella bombylans*),
    127, 128–29
bumblebee.org (website), 69
bumblebees
    annual colonies, 6
    climate change effects on, 235–37
    courtship and mating behaviour, 67–69
    cuckoo bees of, 107–11
    daughter queens, 27
    decline in, 9
    diversity of species, xvi
    first spring sighting of, 19–20

— 267 —

bumblebees (*continued*)
  hibernation, 249–250
  hoverfly mimicry of, 127–29, 156
  laying of fertilised eggs, 25–27
  laying of unfertilised eggs, 27
  male, 27, 67–72
  neonicotinoids concerns, 99
  nest life span, 28
  pollen and nectar collection, 25
  pollen baskets, 36, 65
  sex differences, 65–66
  shivering of flight muscles, 26, 51
  siting of nests, 24–25
  social nature of, 6
  sonication, 96
  species of, 19
  stinging by, 120–21
  tomato cross-pollination, 2
  winter survival of queens, 19
  See also specific types
Burrell, Charlie, 192–93
Bury Litton, 177
Busch, Joseph, 74
Butterfly Conservation, 130, 185
butterfly sightings, 224
buzzing sounds of bees, 95–97
buzz pollination (sonication), 96
BWARS. *See* Bees, Wasps and Ants
  Recording Society (BWARS)

C
cabin by the stream, 91–93, 97
cacao tree, pollination by tiny biting
  midges, 130
Caledonian Forest visit, 134–39
California, almond agriculture, 2–3, 15
Callanish Stones, 145
camouflage vs. mimicry, 127–28
Cane, James H., 183, 184–85
carbon sinks, peatlands as, 144–45
carder bee mimic (*Sericomyia superbiens*), 128
carding behaviour, 198, 199
carpenter bees, 38
Carson, Rachel, 98, 101

catkin-producing trees, 20, 178–79
cavity-nesting solitary bees
  attracting to the garden, 40–43
  life cycle of, 40
  at Malvern Hills, 247
  nest-building behaviour, 37–39
  stinging behaviour, 121
  See also specific types
Center for Food Safety, 98
Ceratina spp., 38
Chamaenerion angustifolium (rosebay
  willowherb), 240–41
Chandler, Phil, 6, 10
China, Sichuan region, apple production,
  179–180
Chipko movement, 175
Chloris chloris (greenfinch), 217
citizen science, 185
Clarke's Mining bee (*Andrena clarkella*)
  nest-building behaviour, 37, 38
  willow pollen collection, 36, 40
classification, biological, 53–56, 55*t*
cleptoparasites (brood parasites), 42–43, 104
  See also cuckoo bees
Clifford, Sue, 176, 177, 217
climate change
  drought-tolerant plants, 193
  Great Yellow bumblebee decline and, 152
  late or early emergence of bees and,
    232–38
  need for year-round succession of
    blooming plants, 24
  urgency of problems, 234
clover, 155
cob bricks, 232
Colletes hederae. See Ivy bee
  (*Colletes hederae*)
colonies
  annual, 6
  unmanaged, 3–4
  wild, 114–16
colony collapse disorder (CCD)
  factors in, 14
  increased interest in beekeeping and, 9

L

lacewings, 126

ladybirds, 219–220, *219*, 221

lamb's ears (*Stachys byzantina*), *191*, 193–95, 196–98, 200, 201, 202

Lamiaceae (mint family), 195

Large bee-fly (*Bombylius major*), 129

Large Sallow Mining bee (*Andrena apicata*), 247

larvae

  bee-flies, 129

  bumblebees, 26, 29

  cuckoo bees, 104, 110

  honeybees, 13–14

  solitary bees, 40

  Stylops spp., 61

  wasps, 34

  wax moths, 29–30

  Wool Carder bees, 199, 201

Lasioglossum spp., 121

leaf-cutter bees

  lining and sealing of nests, 43

  nesting box use, 42, 43

  origin of name, 40

  stinging behaviour, 121

left-alone places, 22

leg posturing, *Macropis* female bees, 183–84

Lesley (beekeeper), 7

Leucozona lucorum (Blotch-winged hoverfly), 128

life cycles

  bumblebees, 24–28

  cavity-nesting bees, 40

  climate change-related concerns, 232–38

  cuckoo bumblebees, 109–10

  Early bumblebees, 63, 64, 71

  Great Yellow bumblebees, 151

  ground-nesting solitary bees, 40

  honeybees, 13

  Ivy bees, 217–18, 225–26

  reproductives, 64

  solitary bees, 67

  Tree bumblebees, 118–19

The Life of the Bee (Maeterlinck), 4

Linnaean taxonomy, 54–56, 55*t*

Linnaeus, Carl, 54

Loch Garten, 134–35

log hives, *1*, 11–12

Long-horned bee (*Eucera longicornis*), 132

Louise (friend), 240–251

lousewort, 143–44

Lwin, Kathryn, 210

Lycaena phlaeas (Small Copper butterfly), 224

Lychnis flos-jovis (flower-of-Jove), 198

Lysimachia spp., 183, 185, 186

M

MacDonald, Archie, 169–170

machair (coastal habitat), 133–34, 148–49, 153–55

Macropis europaea (Yellow Loosestrife bee), 104, 182–88

Macropis nuda, 183, 186

Maeterlinck, Maurice, 4, 5

magnifying glasses, 223

male bumblebees

  courtship and mating behaviour, 19, 67–69

  cuckoo bees of, 110

  Early bumblebee, 63–65

  identification of, 65

  from unfertilised eggs, 27

male honeybees. *See* drones

male solitary bees

  courtship and mating behaviour, 66–67

  Hairy-footed Flower bees, 229–230

  Wool Carder bees, 196–97, 201

Malvern Hills

  awareness of nature on, 249–40

  bee sightings, 247–48

  birch and rowan trees, 242–43, 248–49, 251

  childhood collections from, 173

  commuting walks, xiv

  hawthorn hedges, 243–44

  holly, 242

ABOUT THE AUTHOR

Charlotte Strawbridge

B rigit Strawbridge Howard is a bee advocate, wildlife gardener and naturalist. She writes, speaks, and campaigns to raise awareness of the importance of native wild bees and other pollinating insects. Brigit lives in North Dorset with her husband, Rob.

the politics and practice of sustainable living

CHELSEA GREEN PUBLISHING

Chelsea Green Publishing sees books as tools for effecting cultural change and seeks to empower citizens to participate in reclaiming our global commons and become its impassioned stewards. If you enjoyed reading *Dancing with Bees*, please consider these other great books related to nature and biodiversity.

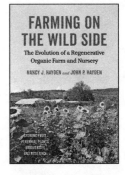

EAGER
The Surprising, Secret Life of Beavers and Why They Matter
BEN GOLDFARB
9781603589086
Paperback • $17.95

FARMING ON THE WILD SIDE
The Evolution of a Regenerative Organic Farm and Nursery
NANCY J. HAYDEN and JOHN P. HAYDEN
9781603588287
Paperback • $29.95

LANDFILL
Notes on Gull Watching and Trash Picking in the Anthropocene
TIM DEE
9781603589093
Hardcover • $25.00

TOP-BAR BEEKEEPING
Organic Practices for Honeybee Health
LES CROWDER and HEATHER HARRELL
9781603584616
Paperback • $24.95

the politics and practice of sustainable living

For more information or to request a catalog, visit **www.chelseagreen.com** or call toll-free **(800) 639-4099**.

Red tailed Bumblebees
Bombus lapidarius on Knapweed
Forrabury Stitches, Boscastle   sunny